T0302657

Developing Headquarters Guidance for Army Installation Sustainability Plans in 2007

Beth E. Lachman, Ellen M. Pint, Gary Cecchine, Kimberly Colloton

Prepared for the United States Army
Approved for public release; distribution unlimited

RAND ARROYO CENTER

The research described in this report was sponsored by the United States Army under Contract No. W74V8H-06-C-0001.

Library of Congress Cataloging-in-Publication Data

Developing headquarters guidance for Army installation sustainability plans in 2007 / Beth E. Lachman ... [et al.].
p. cm.
Includes bibliographical references.
ISBN 978-0-8330-4707-6 (pbk. : alk. paper)
1. United States. Army—Facilities—Management. 2. United States. Army—Facilities—Maintenance and repair. I. Lachman, Beth E., date.

UC403.D48 2010
355.7068'4—dc22

2009045741

Cover photo of Fort Carson Solar Array.

Published 2009 by the RAND Corporation
1776 Main Street, P.O. Box 2138, Santa Monica, CA 90407-2138
1200 South Hayes Street, Arlington, VA 22202-5050
4570 Fifth Avenue, Suite 600, Pittsburgh, PA 15213-2665
RAND URL: http://www.rand.org/
To order RAND documents or to obtain additional information, contact
Distribution Services: Telephone: (310) 451-7002;
Fax: (310) 451-6915; Email: order@rand.org

Preface

Given the experiences of industry and communities, many Army installations have started to develop and implement installation sustainability plans (ISPs). Such plans typically document long-range plans addressing mission, community, and environmental issues developed through a strategic planning process. However, these plans are inconsistent in their focus and scope, and most do not fully address some key sustainability issues. Thus, the Army asked RAND Arroyo Center to conduct a study, entitled "Developing Guidance for Installation Sustainability Plans," to examine Army installations' experiences with developing and documenting these plans, and to recommend Headquarters, Department of the Army guidance and policy to help foster improved development and implementation of installation sustainability plans throughout the Army. This report documents the study results, providing background information on sustainability, describing the installation sustainability planning process, noting progress in ISP implementation, and recommending approaches to improve the ISP process.

This report should be of interest to Army and other Department of Defense staff involved in installation planning and management and sustainability, including mission, environmental, community, and quality of life concerns. It should also be of interest to other federal agencies, state and local governments, nongovernmental organizations, and businesses that interact with Army installations and those that have more general interests in sustainability and installation planning.

This research was sponsored by L. Jerry Hansen, the Deputy Assistant Secretary of the Army, Strategic Infrastructure and Senior Official Performing Duties as Assistant Secretary of the Army, Installations and Environment, and Craig E. College, the Deputy Assistant Chief of Staff for Installation Management. It was conducted within RAND Arroyo Center's Military Logistics Program. RAND Arroyo Center, part of the RAND Corporation, is the U.S. Army's federally funded research and development center for policy studies and analyses.

The Project Unique Identification Code (PUIC) for the project that produced this document is SAIED07268.

For more information on RAND Arroyo Center, contact the Director of Operations (telephone 310-393-0411, extension 6419; FAX 310-451-6952; email Marcy_Agmon@rand.org), or visit Arroyo's web site at http://www.rand.org/ard/. This report can be found on the web at http://www.rand.org/pubs/monographs/MG837/.

Contents

Figures

Tables

Boxes

Summary

Background and Purpose

As is the case with businesses and governments at all levels, the Army has recognized the need to manage its installations in a way that sustains them for the future. Based on lessons from industry and communities, many Army installations started to develop installation sustainability plans (ISPs). Beginning at the turn of the century, a few installations, like Fort Bragg, started developing and implementing installation sustainability plans because of the operational, financial, and environmental benefits they saw industry and communities experiencing by implementing sustainability approaches. Such plans address long-range mission, community, and environmental issues and priorities and are developed through a strategic planning process. To develop an installation sustainability plan, installation staff members and key stakeholders define the vision and goals for the installation over a 20- to 25-year horizon and identify tasks needed to achieve them. Such plans are often developed in addition or in combination with installation strategic plans.

The installations that have developed sustainability plans have provided insight to the rest of the Army about how to develop such plans. The plans have also been useful to installation management and staff. However in 2007, we found that these plans were inconsistent in their focus and scope, and most did not fully address some key sustainability issues, such as quality of life (QOL) and master planning issues. Nor was there a formal Headquarters, Department of the Army (HQDA) policy requiring installation sustainability planning, and no HQDA guidance describing specific issues that should be addressed by the plans or guidance on how to develop or implement an ISP.

This study aimed to develop guidance for ISPs to ensure that they strategically and comprehensively deal with installation sustainability issues affecting mission, community, and environment. Additionally, the guidance should foster the effective development and implementation of ISPs throughout the Army. As part of this study, we examined the ISP development process and implementation progress, challenges, and needs to ensure that the guidance enabled effective ISP development and implementation. The methodology for the study included examining ISP documentation; reviewing sustainability and management literature; interviewing installation staff, Army regional, and HQ staff, and other relevant experts; attending ISP development workshops and visiting several installations that have implemented ISPs; and assessing all this information. This study was conducted from fall 2006 through fall 2007. It is important to note that there have been many changes to the program since this research was conducted that address many of the issues raised in this study. Some examples have been provided in this document. For more current information, contact the Assistant for Sustainability, Office of the Deputy Assistant Secretary of the Army Environment, Safety and Occupational Health (DASA-ESOH) or the Assistant Chief of Staff for Installation Management's Environmental Division (DAIM-ISE).

Findings and Recommendations

Our study found that some installations, such as Forts Bragg, Carson, Hood, and Lewis, have made considerable progress in developing and implementing ISPs with limited resources and guidance. For example, Fort Bragg built an urban village training site in 90 days from 50 recycled shipping containers at an estimated savings of $220,000 compared to standard construction and has built brigade complexes to be more aesthetic and friendly to pedestrians and the environment. Fort Hood saved more than $2.5 million in 2006 through its qualified recycling program, compost recycle program, inert material management, deconstruction management, special waste management, and electronics waste recycling program. Fort Hood also used recycled tires to create a platform for a tank firing range to reduce dust and air-quality impacts. Fort Lewis has pioneered the Sustainable Interiors Showroom approach that has resulted in the purchase of more than $180,000 in recyclable and/or recycled content furnishings on post, and purchased 10 percent green power in FY07. Energy initiatives saved Fort Carson approximately $1 million in FY06, and this post is participating in a strategic ecoregion management collaboration, called the Central Shortgrass Prairie Ecoregion Partnership, to help prevent species-protection requirements from causing training restrictions.

However, we also found that issues remain. For example, most of the accomplishments have been in environmental technology and pollution-prevention areas and range management, maintenance, and use. Yet key sustainability areas—such as family support, health, and other QOL issues; regional transportation and growth management; ecosystem management; and other mission functions—receive less attention. As a result, there are fewer accomplishments in ISP processes for these areas.[1] In addition, installation sustainability project accomplishments are not being analyzed or documented well, making it difficult to transfer lessons to other installations, to measure and track true progress and benefits, and to ensure that programs are addressing the

---

[1] It is important to note that installations that have implemented ISPs may have accomplishments in these other areas through other programs, but they have not yet been integrated into the ISP process, as is discussed in greater detail in this report. Also, much more has been done since 2007 to address such areas in the ISP process.

most pressing sustainability needs. Limited numbers of staff with sustainability expertise and funding constraints make it difficult to develop and implement ISPs. Often the funding for ISP implementation comes through environmental programs, which is one reason why it is difficult to focus on QOL and other nonenvironmental sustainability issues. A lack of mechanisms to benefit from sustainability activities return on investments also limits ISP implementation. Army policy and standards, such as U.S. Army Corps of Engineers (USACE) building design standards, can also impede sustainability implementation and progress.

To address these and other issues identified, we had two sets of recommendations. One pertains to the development of ISPs and the other to their implementation.

Development of ISPs

By early FY07, ISPs were being developed through a series of workshops held at each installation. This process had both advantages and disadvantages. The advantages included increasing the education and enhancing the motivation of installation staff with respect to improving sustainability and the potential for developing a high-level advocate for sustainability issues.

The disadvantages were that the workshops were time-consuming and expensive. It often took over a year to hold the required sessions, and they cost in total about $140,000 per installation. They also tended to limit the number of installations that can develop ISPs, because the Army only had staffing and resources for about four to eight sets of workshops a year. Nor did they necessarily address all the key issues: the workshops tended to focus mostly on sustainability examples and objectives with obvious economic benefits. While financial returns are important, so too are regional sustainability and ecosystem concerns, which could have important long-term implications. Finally, there was no requirement that installations complete or implement an ISP after the workshop process ended, so a couple of installations have gone through the workshop process but have not developed a written document or implemented a plan.

We recommended three main improvements for the workshop process. First, it should be streamlined so that more installations can develop ISPs in a more timely fashion. One workshop should be sufficient to initiate the process and to garner the benefits of the workshop process. In addition, this workshop's sustainability presentations need to be customized to focus more on the unique mission, quality of life, community, and environmental concerns of each installation given its local circumstances. Note that since this research was done in 2007, ACSIM has streamlined the workshop process as suggested here.

Second, installations need more technical support to develop an ISP document once the workshop process is completed.

Third, there should be sufficient guidance and support to ensure that key functional staff and stakeholders are involved in ISP development, and that such functional areas as health, education, and morale, welfare, and recreation are considered in the sustainability planning process. This guidance should include standard reporting requirements for the contents of ISPs and periodic progress reports.

Implementation of ISPs

With respect to implementation, we had two recommendations. First, the Army should provide HQDA-level implementation guidance. Second, it should broaden the involvement of Army organizations outside the installation management and environmental organizations currently participating in the ISP implementation process. We discuss each recommendation in turn.

HQDA-Level Guidance. First, the Army should direct garrison commanders and other senior installation staff to support sustainability, including investing staff time, funding, and other resources. This includes designating a staff member as the sustainability coordinator and providing that person the time—at least 50 percent of their work hours—to carry out these duties, at least until a sustainability program is well established. In addition, sustainability should be included in garrison commander and other senior installation staff performance evaluations, education, and training.

The Army also should address problems with funding sustainability initiatives, because they may require an upfront investment to achieve the expected life-cycle cost reductions or operational benefits. To help integrate sustainability practices into installation operations, HQDA should direct the inclusion of sustainability issues in key installation strategic planning, implementation, and operational documents, such as installation strategic and master plans and range management plans. It is important to note that this issue was partly addressed in a December 2007 update of Army Regulation (AR) 200-1 that directs installation commanders to "Ensure that the installation strategic planning office (or equivalent) incorporates sustainability principles into strategic and other installation management plans."[2] In addition, combining the ISP and installation strategic plan can be an effective way to start to integrate sustainability into key installation business processes. Since late in 2007, the ACSIM-funded ISPs have incorporated sustainability planning into the installation strategic plan.

Additionally, the Army should ensure that installations focus more attention on gaps in implementation, including QOL issues, regional collaboration, ecosystem management and other strategic natural resource management activities, and mission areas other than ranges. Regional collaboration and approaches are especially needed in such strategic areas as growth management, encroachment issues, and natural resource constraints that are likely to arise in the future. Since late 2007, such issues are being addressed more routinely in ISPs.

---

2 U.S. Army Regulation 200-1, "Environmental Protection and Enhancement," December 13, 2007, addresses environmental responsibilities for all Army organizations and agencies.

The Army also needs to provide a channel that enables installations to identify and communicate policy and funding issues that hinder sustainability implementation. Installations that identify a policy roadblock, such as an Army regulation, should be able to flag it to the appropriate HQDA organization, which should then ensure that the relevant policy is changed or a new process is put into place.

Next, HQDA should help installations measure the true costs and benefits of sustainability projects. It is difficult for installations by themselves to do such assessments given the cost and skills required. Further, many projects would be beneficial across a wide range of Army installations. Therefore, Army headquarters and regional organizations should provide resources to conduct some initial assessments of existing or new sustainability projects and document them so the value can be measured and lessons can be transferred. The Army should also ensure that information about successful sustainability projects is shared across installations by providing support for documenting and transferring lessons learned.

Finally, the Army should endeavor to change its culture through sustainability guidance so that sustainability is viewed as a long-term strategic Army-wide program, not just an environmental one. The 2004 "Army Strategy for the Environment," which provided the Army's definition of sustainability, along with ongoing activities to integrate sustainability into all parts of the Army are good first steps, but sustainability needs to be implemented across the Army. Sustainability should be viewed as being as important as safety, for example. The way the Army integrated safety concerns throughout its installations is a good model.

Involving a Wider Range of Army Organizations. Many organizations that are not currently or directly involved in installation sustainability could help with ISP implementation. The Office of the Army Deputy Chief of Staff, G-3/5/7, and U.S. Army Training and Doctrine Command (TRADOC) should include sustainability training in unit and garrison pre-command courses and garrison staff education. The Army should integrate sustainability into Army-wide guidance and standards that affect installation operations, including USACE standards of excellence for building and range design guides (such as *Training Circular 25-8: Guidance for Ranges*), Integrated Training Area Management guidance, and budgeting, programming, and procurement guidelines. Base Realignment and Closure (BRAC) transformation and implementation guidance also should mention the importance and need to follow installation sustainability practices. HQDA guidance on compatible buffers and natural resource management plans should also include sustainability.

Army headquarters organizations and commands, such as G-3/5/7, USACE, and nonenvironmental functional organizations within the Installation Management Command (IMCOM) and the office of the Assistant Chief of Staff for Installation Management (ACSIM), should provide more support for sustainability. Such support should ensure that their guidance, policies, practices, and training include sustainability elements. It can also include providing needed funding, such as sufficient military construction funds to enable building projects to meet sustainability standards.

Installations have accomplished much in developing and implementing ISPs with limited resources and guidance. Given such accomplishments and the experience from sustainable community and industry activities, we conclude that successful ISP implementation can significantly benefit Army missions, quality of life, and the environment. Additional support along the lines of the recommendations of this report will make installation sustainability actions even more effective and are needed if the Army expects installations Army-wide to develop and implement ISPs.

Acknowledgments

We would like to thank L. Jerry Hansen, the Deputy Assistant Secretary of the Army, Strategic Infrastructure and Senior Official Performing Duties as Assistant Secretary of the Army, Installations and Environment, and Craig E. College, Deputy Assistant Chief of Staff for Installation Management, for sponsoring this research. We also thank members of their staffs, including Chien Huo, Brian Sangtinette, COL Jeffrey Phillips, and COL Michael Bennett, for providing information and feedback on our emerging results. We also appreciate the help of staff members with whom we spoke at other Army organizations, including the Army Environmental Policy Institute, the Office of the Assistant Secretary of the Army for Installations and Environment, the Installation Management Command, the U.S. Army Corps of Engineers, and about a dozen installations. A special thanks goes to installation staff members who hosted our visits and allowed us to attend their sustainability workshops, including Forts Bragg, Carson, and Hood, U.S. Army Garrison Hawaii, and the Pennsylvania National Guard.

This report also benefited from reviews and comments by several RAND colleagues, including formal reviews by David Oaks and Susan Resetar. We thank Jerry Sollinger for helping with the structure, organization, and presentation of material within this report.

List of Acronyms

ACSIM	Assistant Chief of Staff for Installation Management
ACUB	Army Compatible Use Buffer
AEPI	Army Environmental Policy Institute
AR	Army Regulation
BCTC	Battle Command Training Center
BRAC	Base Realignment and Closure
CERL	Construction Engineering and Research Lab
DOIM	Directorate of Information Management
DPTM	Directorate of Plans, Training, and Mobilization
DPW	Directorate of Public Works
DRM	Directorate of Resource Management
EB	Existing building
EMS	Environmental management system
FORSCOM	U.S. Army Forces Command
FY	Fiscal year
GEMI	Global Environmental Management Initiative
GSA	General Services Administration
HQDA	Headquarters, Department of the Army
IDG	Installation design guide
IMCOM	Installation Management Command
INRMP	Integrated Natural Resource Management Plan
ISP	Installation sustainability plan
ITAM	Integrated Training Area Management
LEED	Leadership in Energy and Environmental Design
LEED-EB	LEED for existing buildings

LEED-NC	LEED for new construction
LID	Low-impact development
LLP	Long-leaf pine
MEDDAC	U.S. Army Medical Department Activity
MWR	Morale, Welfare, and Recreation
NB	New building
O&M	Operations and maintenance
P2	Pollution prevention
PAIO	Planning, Analysis, and Integration Office
PPSIP	Pikes Peak Sustainability Indicators Project
QOL	Quality of life
ROI	Return on investment
SIS	Sustainable Interiors Showroom
SPiRiT	Sustainable Project Rating Tool
SSA	Strategic Sustainability Assessment
T&ES	Threatened and endangered species
TRADOC	U.S. Army Training and Doctrine Command
USACE	U.S. Army Corps of Engineers
USAG	U.S. Army Garrison
USGBC	The U.S. Green Building Council
WBCSD	World Business Council for Sustainable Development
WDFW	Washington Department of Fish and Wildlife
WDNR	Washington Department of Natural Resources

CHAPTER ONE

Introduction

Since the 1990s, sustainable development—a long-term systems approach that addresses environmental, economic, and social issues in planning and development through integrated systems and holistic thinking—has been a key concept for infrastructure, economic, social, and environmental planning, development, and management in communities, businesses, and countries throughout the world. Many organizations in the Army, such as U.S. Army Forces Command (FORSCOM) and individual installations, recognized the advantages of sustainable development and began to apply aspects of this approach to their installation planning in the early 2000s. Based on lessons from industry and communities employing sustainability practices, installations started to develop installation sustainability plans (ISPs). Since then a number of Army installations have developed ISPs, including Fort Bragg, Fort Carson, Fort Hood, Fort Lewis, Fort Stewart, and Fort Benning.

An ISP, developed through a strategic planning process, addresses long-range mission, community, and environmental priorities and issues. To develop one, cross-functional teams from the installation, working under the guidance of the commander and in coordination with regional stakeholders, define the vision and goals for the installation over a 20- to 25-year horizon and identify tasks needed to achieve them.

The initial installations that have developed and implemented ISPs have served as excellent "test beds" for Army best practices with respect to sustainability. The success achieved by these installations and their lessons learned served as the foundation for the goals of the 2004 "Army Strategy for the Environment," which focuses on sustaining Army operations far into the future by strategically addressing the interrelationships of mission, environmental, and community concerns.[1] However, there has been inconsistency from installation to installation in the focus and scope of the plans, and most were not yet fully addressing some key issues in sustainability, such as quality of life and master planning. In March 2006, the Installation Management Command's Southeast Region released an informal guide on installation sustainability planning, which focuses mostly on how to conduct the planning process and provides

[1] U.S. Army, 2004.

information about sustainability. However, in 2007 there was no formal Headquarters, Department of the Army (HQDA) policy requiring installation sustainability planning, no HQDA guidance that describes specific issues that should be addressed by the plans, and no guidance on how to develop or implement an ISP.

The purpose of this study was to develop guidance for installation sustainability plans to ensure that they strategically and comprehensively address installation sustainability issues affecting mission, community, and environment. Such guidance would ensure that key issues that had received less focus in the process, such as master planning, quality of life, and potential synergies across installations, were also addressed in the process. The study focused on developing recommended guidance that would help foster both the effective development and implementation of installation sustainability plans throughout the Army. Implementation progress, challenges, and needs were examined to ensure the guidance enabled effective implementation.

Definition of Sustainability

The "Army Strategy for the Environment"[2] describes sustainability as its foundation. It states that "a sustainable Army simultaneously meets current as well as future mission requirements worldwide, safeguards human health, improves quality of life, and enhances the natural environment." This strategy transitions the Army's historically compliance-based environmental program to a mission-oriented approach based on the principles of sustainability, and it builds on the success of the Army's strategic "beyond compliance" activities in areas such as pollution prevention and ecosystem management. Sustainability requires systems thinking that integrates mission, environment, and community issues. These three types of issues are referred to as the "triple bottom line." The strategy recognizes that integrated strategic system approaches for sustainable operations, installations, systems, and communities will enable the Army mission over the long term more effectively than current practice. Sustainability planning seeks approaches that simultaneously satisfy mission, environmental, and community needs through innovative systems thinking while also trying to reduce long-term costs, including waste treatment and disposal costs, facility operating and maintenance costs, and energy costs.

Outside the Army, there is no consistent definition of sustainability, although most definitions focus on long-term, integrated approaches to social, economic, and environmental issues. The most commonly used definition was established by the United Nations' World Commission on Environment and Development (the Brundtland Commission) in its 1987 report, *Our Common Future*.[3] It defines sustainable

2 U.S. Army, 2004.

3 World Commission on Environment and Development, 1987.

development as "development that meets the needs of the present without compromising the ability of future generations to meet their own needs."

Given the economic, marketing, societal, and environmental benefits, many businesses have embraced sustainable development for their environmental programs and in strategic planning. Companies perceive potential financial and operational benefits from implementing more-sustainable practices. These include reducing costs and liabilities, increasing customer loyalty and market position, protecting businesses' right to operate, and developing new products.[4] Recognizing such benefits, many companies have become active in the international sustainable development agenda. The World Business Council for Sustainable Development (WBCSD) is a consortium of about 200 international companies sharing a commitment to the environment and to the principles of economic growth and sustainable development. These companies include such large corporations as AT&T, 3M, Arthur D. Little, DuPont, Dow Chemical, Eastman Kodak, General Motors, Nissan, Mitsubishi, NEC, Johnson & Johnson, P&G, Seiko Group, Shell International, Weyerhaeuser, Toyota, and AOL Time Warner. These companies are trying to evolve toward sustainability and develop their own definitions and strategies.[5]

Many community-based efforts use the term sustainable community, which emphasizes the community aspect of sustainable development. A "sustainable community" is usually defined uniquely by each community, based on its interests, needs, and culture. It typically involves a long-term, integrated, systems approach to developing and achieving a healthy, enduring community by jointly addressing economic, environmental, and social issues. Building consensus and fostering partnership among key stakeholders about community problems and solutions is also important to such efforts.[6]

Figure 1.1 shows the relationships between various types of environmental sustainability efforts. This simplified figure illustrates relationships between different activities related to sustainability and the ultimate goal of sustainable development and a sustainable earth.[7] First, the focus and interests of the traditional industry and technology experts are presented on the left side of the figure. At an operational level, such techniques and policies as pollution prevention (P2), design for the environment (known as DfE), environmental management systems (EMSs), and environmental

---

[4] This list of benefits was based on hundreds of interviews and discussions with business people regarding sustainable development in a World Resources Institute (WRI) study: Arnold and Day, 1998.

[5] For more about businesses embracing sustainability and their definitions, see the World Business Council for Sustainable Development web site at http://www.wbcsd.org/.

[6] See, for example, Lachman, 1997, for a more detailed discussion of the concept of a sustainable community.

[7] While "sustainable development" often means meeting current needs without compromising future needs, "sustainable earth" refers to the idealistic goal that sustainable development has been achieved everywhere on the earth.

Figure 1.1
The Relationship of Sustainability Efforts

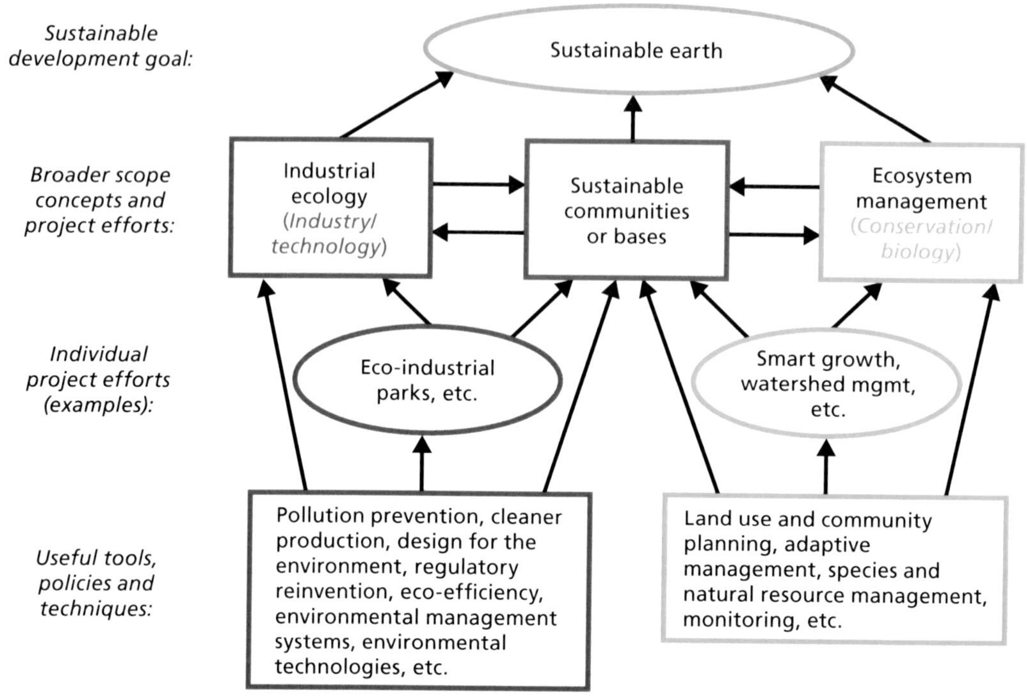

SOURCE: Lachman et al., 2001.
RAND *MG837-1.1*

technologies are implemented. Such tools are used in individual projects, such as company environmental projects and eco-industrial parks, which group manufacturing and service businesses to jointly manage environmental issues and resources, including energy, water, and materials. For example, waste products or waste heat generated by one process can be used by other businesses in the group. Such efforts contribute to the development of broader concepts and efforts toward industrial ecology, which involves shifting industrial processes from "open loop" systems that generate waste to "closed loop" systems where waste products become inputs for other processes. Given the traditional interests and needs of manufacturing and industrial facilities, it makes sense that they would mainly focus on technology issues.

Next, on the right-hand side of Figure 1.1, are the traditional views of natural resource and land managers, who tend to focus on biological, land-use planning, and conservation issues. At an operational level, techniques and policies used here include land-use planning, adaptive management, and species and natural resource management. Individual project efforts focus on watershed management, smart growth/growth management plans;[8] and individual preserve, wilderness, and park management. Such efforts contribute to the development of broader concepts and efforts toward ecosystem management.[9]

This figure shows the disconnects between some of the industry and technical approaches and the land-use planning and natural resource approaches. These disconnects often result from traditional disciplinary ways of thinking and orientation and from questions about who has primary authority in the efforts. Industry managers and engineers traditionally focus on technology and economic issues. Natural resource and land managers and biologists tend to focus on conservation and land management issues related to flora and fauna. This difference is the classic environmental education split between the "technology-engineering" experts and the "bugs and bunnies" experts.

However, environmental approaches have been changing. In practice, more interaction and integration are starting to take place across these areas than Figure 1.1 suggests. However, such interactions are still not the norm, except in one main area: sustainable community activities. Both sides of the figure have been integrated in various sustainable community activities. At the community level, all of these concerns and issues regarding industry, technology, land management, and conservation come into play. Unlike most industries and most public natural resource management activities, U.S. communities include industrial, commercial, natural resource management, and residential activities. To bridge the disconnects this figure illustrates, communities are trying to break out of traditional disciplinary, stovepipe, and media approaches[10] for their sustainability efforts.

Army installations are often very much like communities, rather than most businesses, in their functions and activities and could replace sustainable communities in Figure 1.1. Army installations' environmental management activities have to balance and plan for industrial, commercial, natural resource management, and residential

---

[8] "Growth management" and "smart growth" refer to a planning and administrative approach that focuses on supporting and coordinating the development process. The concept is oriented toward guiding community development rather than restricting growth. Most growth management plans of local, state, and regional governments are focused on accommodating development while maintaining communities' quality of life, economic base, and environmental qualities. A practical definition of growth management is "a dynamic process in which governments anticipate and seek to accommodate community development in ways that balance competing land use goals and coordinate local and regional interests" (Porter, 1997).

[9] Place-based management efforts related to natural resource management often focus on ecosystem management. Definitions of "ecosystem management" also differ, but the following is well accepted in the scientific community: "Ecosystem management integrates scientific knowledge of ecological relationships within a complex sociopolitical and value framework toward the general goal of protecting native ecosystem integrity over the long term" (Grumbine, 1994).

[10] "Media approaches" means approaches that focus on environmental media, such as air, water, hazardous substances, and flora and fauna. Such approaches include traditional air, water, hazardous substances, and species compliance and management activities.

activities. As a result, installation sustainability is more likely to resemble community sustainability than corporate sustainability. However, Army installations often have stovepiped responsibilities and disciplinary disconnects between environmental engineering, land-use planning, and natural resource staffs, as will be discussed later in this report.

Methodology

In order to understand the Army's experience with developing and implementing installation sustainability plans and develop guidance for improving these processes, the research team carried out several activities.

First, we obtained and reviewed many of the existing ISPs, as well as a guide to installation sustainability planning produced by the Installation Management Command (IMCOM) Southeast Region. We also examined the literature on industry and community sustainability practices to identify guidance and approaches that are relevant for the Army. Relevant business and environmental management literature was also reviewed.

Second, we interviewed Army and external experts on sustainability, including representatives of the Assistant Chief of Staff for Installation Management (ACSIM) Office of the Director of Environmental Programs (ODEP), the Army Environmental Policy Institute (AEPI), the IMCOM Southeast Region, and consultants employed by the Army to assist with installation sustainability planning.

Third, we visited some of the Army installations that have ISPs in place, including Forts Bragg, Carson, and Hood. We also conducted telephone interviews with people at most of the other installations that had developed or were developing ISPs by 2007, including Forts A.P. Hill, Benning, Campbell, Eustis, Jackson, Lewis, and Stewart. At each installation, we interviewed participants in the sustainability planning and implementation process. We also reviewed installation, Army, journal articles, public press, and other documentation regarding sustainability accomplishments at these installations. We used this information to help understand the ISP processes and review the progress in implementing the ISP at each installation. We also attended ISP development workshops held during fiscal year 2007 (FY07) for the U.S. Army Garrison (USAG) Hawaii and the Pennsylvania National Guard.

Based on the information gathered, we assessed and identified areas where progress has been made in implementing ISPs, gaps in progress, and the barriers to achieving progress. We used these findings to develop recommendations for areas where HQDA guidance or other actions could help installations develop and implement ISPs.

This research was conducted from fall 2006 through fall 2007, and since that time there have been some changes to the Army's ISP program, including implementing some of the recommendations discussed in this report.

Outline of This Report

In the remainder of this report, Chapter Two provides background information on the development and implementation of ISPs. Chapter Three reviews the progress that installations have made in implementing sustainability. Chapter Four reports on our assessment of the needs of the ISP development and implementation processes, including what is needed to improve these processes. Chapter Five discusses recommendations to improve the ISP development and implementation processes.

CHAPTER TWO

Development and Implementation of ISPs

In this chapter we review the Army's experience with developing and implementing installation sustainability plans. We define an ISP and then describe the early 2007 guidance on ISPs, the process that was used to develop ISPs by FY07, what is typically included in an ISP, and how ISPs are being implemented.

What Is an ISP?

An installation sustainability plan addresses long-range mission, community, and environmental issues and priorities. It is developed through a strategic planning process that brings together representatives of the major installation support directorates and other activities, such as the Directorate of Public Works (DPW), the Directorate of Plans, Training, and Mobilization (DPTM), the Directorate of Resource Management (DRM), the Planning, Analysis, and Integration Office (PAIO), and the contracting office. The process may also include representatives of tenant units and other tenant activities, family readiness groups, Installation Management Command (IMCOM), the U.S. Army Corps of Engineers (USACE), the General Services Administration (GSA), and other organizations whose decisions or policies affect installation sustainability. The planning process also often involves stakeholders from the surrounding community.

The participants form teams that may be organized based on major installation or management functions (military training, infrastructure development and maintenance, procurement, transportation, etc.) and/or subject areas (water, air quality, energy, lands, materials, etc.). Each team defines a long-term vision and goals for the installation over a 20- to 25-year time horizon, and then identifies a set of tasks that can be accomplished over the next few years to make progress toward the goals. For example, Fort Carson's long-term sustainability goals include reducing total water purchased from outside sources by 75 percent; reducing the total weight of hazardous air pollutant emissions to zero; and bringing all applicable facilities up to the Leadership in Energy and Environmental Design (LEED) Platinum Standard or higher.[1] Shorter-term activities undertaken to help meet each of these goals included installing rain sensors on irrigation systems, changing the types of paint used for vehicles and road striping to reduce hazardous air pollutants, and requiring all major new construction projects to meet the LEED Gold Standard.[2]

In order to secure broad, high-level participation in the planning process, usually the garrison commander supports and encourages the process. The garrison commander also participates in the ISP development process by providing feedback to the teams and prioritizing goals.

ISPs were often developed separately from other installation planning processes to keep the focus on sustainability, but they can integrate parts of the other plans into the sustainability process and vice versa, such as parts of the Integrated Natural Resource Management Plan (INRMP) being integrated into the ISP. By early 2007, most ISPs were being developed separately from installation strategic plans because they were being driven by installation staff focused on implementation, such as environmental management staff, rather than PAIO staff that focuses on strategic planning. A separate ISP process has been important to help educate staff about and motivate them to implement sustainability practices. When sustainability practices have been established and installation staff are more familiar with sustainability concepts, ISPs can be successfully integrated with other installation strategic planning processes without losing their unique focus.

Guidance on ISPs by Early 2007

FORSCOM, headquartered at Fort McPherson, Georgia, was one of the early Army proponents of sustainability planning. Several of the installations that first developed ISPs were also located in the southeast region of the United States. When installation management functions were consolidated into the Installation Management Agency (recently renamed the Installation Management Command), much of the Army's sustainability expertise was concentrated at the IMCOM Southeast Regional office. In March 2006, this office released a guide on installation sustainability planning.[3] It was essentially a collection of PowerPoint briefings, with links to other documents and references, describing a four-workshop process to help Army installations develop ISPs and providing background and motivational information about sustainability.

[1] LEED is a voluntary, consensus-based national rating system for developing high-performance, sustainable buildings. It is an industry building standard developed by the U.S. Green Building Council, which also provides a certification process for LEED-rated buildings. For more information about LEED standards, see Appendix A.

[2] Fort Carson, August 2002.

[3] U.S. Army Installation Management Command Southeast Region, 2006.

ACSIM provided funding and some personnel support for seven installations to conduct workshops in 2007 to develop ISPs as defined by the IMCOM Southeast Region guidance. This group included the Pennsylvania National Guard; USAG Hawaii; USAG Wiesbaden, Germany; Letterkenny Army Depot; Fort Detrick, Maryland; and the California National Guard.

To start the sustainability planning process, the Chief of the Sustainability Division in the Office of the Director of Environmental Programs at ACSIM first met with the garrison commander to ensure that he or she will support the effort. The installation then scheduled a series of workshops that guide participants through the development of an installation sustainability plan. These workshops required 15–20 sustainability experts (both Army and contractor personnel) to make presentations and help facilitate the work done by each of the teams. ACSIM had been funding and supporting the process of developing and implementing ISPs, though there was no official headquarters guidance about developing and implementing ISPs.

The Army had a draft document entitled "The Strategic Plan for Army Sustainability" that had not yet been officially released as of December 2007. This 2007 draft strategic plan included some draft guidance about developing ISPs. It would establish a critical task that installations should develop and complete ISPs that include 25-year sustainability goals in collaboration with local communities. The targets for completing this task were: all Army tier 1 installations should develop and be implementing their ISPs by 2010, tier 2 by 2012, and all installations by 2015.[4] Since the draft was a broad strategic plan for promoting sustainability throughout the Army, it would not provide specific guidance on what should be in an ISP, nor how to develop or implement one. However, this might change by the time any such draft guidance becomes official guidance.

In January 2007, the President issued Executive Order 13423, "Strengthening Federal Environmental, Energy, and Transportation Management."[5] This document set several medium-term goals for all federal agencies, including reducing energy intensity by 30 percent by 2015; increasing usage of renewable energy; reducing water consumption by 16 percent by 2015; increasing sustainable procurement of goods and services; incorporating sustainable practices into 15 percent of their capital asset building inventory; and decreasing petroleum usage and increasing non-petroleum-based fuel consumption by vehicle fleets. The ISP process could help the Army and its installations meet these goals.

---

4 U.S. Army, July 25, 2007.

5 Executive Office of the President of the United States, January 26, 2007.

How ISPs Were Developed

The process used to develop an installation sustainability plan has evolved over the last several years as the installations, IMCOM, and ACSIM have gained experience. In this section we describe the processes used by some of the early adopters of sustainability planning, as well as the process prior to summer 2007. It is important to note that this process has been evolving and changing over time, especially during 2007.

Early Adopters

Fort Bragg was the first installation to start a sustainability program in 1999–2000. In 1999, environmental staff at Fort Bragg started a study to identfy the post's "environmental footprint," i.e., to measure the impact of its operations on the natural environment and adjacent community. This study was used to provide information for Fort Bragg's Environmental Sustainability Executive Conference that was held in April 2001. At this conference, personnel from Fort Bragg, FORSCOM, HQDA, local communities, and environmental regulators identified the environmental challenges to Fort Bragg's long-term sustainability and proposed 25-year sustainability goals. During May and June 2001, integrated teams developed initial objectives and tasks to start Fort Bragg moving toward these sustainability goals. This sustainability activity was and is coordinated by the long-term sustainability planning staff of Fort Bragg's Environmental Sustainment Office.[6]

At about the same time, environmental staffs at Fort Carson, Fort Hood, and Fort Lewis were also developing sustainability programs. In 2002, they each had a single sustainability workshop to help develop installation sustainability goals and cross-functional teams for sustainability. As at Fort Bragg, since the sustainability effort was being initiated by environmental staff, most of the sustainability emphasis was on environmental issues. To illustrate this process and focus we briefly describe Fort Hood's and Fort Carson's initial ISP development processes.

Fort Hood's sustainability program was officially established in June 2001. It held a sustainability workshop in June 2002, attended by a large number of installation staff, environmental professionals, and community leaders. Participants organized into five teams and developed Fort Hood's 25-year strategic sustainability goals (see Table 2.1).

The teams continued to meet after the workshop to develop shorter-term objectives and targets to move toward the goals, as well as metrics to measure progress.

Prior to holding a sustainability workshop in September 2002, Fort Carson developed a detailed sustainability baseline report covering descriptive statistics, key issues and challenges, and existing sustainability activities in nine subject areas: water, air quality, energy, transportation, lands, materials, wildlife, noise, and cultural resources.

---

[6] Fort Bragg, February 2004, p. 3-4.

Table 2.1
Fort Hood's Sustainability Teams and Goals

Sustainability Team	Goal
Training lands	Training landscapes managed to support current and future mission requirements while sustaining cultural, natural, and land resources.
Water quality	Provide high-quality potable water and reduce consumption while maintaining mission readiness and quality of life.
Air quality	Fort Hood will actively reduce its impact on regional air quality from all sources.
Infrastructure and energy	All infrastructure and energy systems are planned, designed, constructed, and maintained to be sustainable and secure.
Products and materials	Fort Hood will use sustainable products and services, with active regional involvement, to minimize waste and environmental impact.

SOURCE: Fort Hood, 2005.

Workshop participants included representatives from Fort Carson, HQDA and subordinate headquarters, other local military installations, civic leaders, regulatory agencies, and the community. They organized into six teams of 25–35 people each (community well-being, energy and transportation, facilities and installation, materials, training lands and ranges, and sustainability management system) to develop 25-year sustainability goals. The teams produced a total of 12 goals. For each goal, the teams defined the problem being addressed, the desired end state, metrics, and intermediate objectives.[7]

An important part of this process was support from FORSCOM, especially in helping to transfer ideas about sustainability across these four installations. For example, these FORSCOM installations had monthly sustainability conference calls to help the installations learn from each other. FORSCOM support, including environmental staff knowledgeable about sustainability, enabled the installations' sustainability teams to develop and start implementing their ISPs after the sustainability workshop was over.

Process for Developing ISPs

Over time the ISP development process had evolved to a series of four sustainability workshops designed to help develop the goals, teams, initial objectives, and tasks for the installation sustainability plan based on the IMCOM Southeast Region's "Guide to Creating a Sustainable Installation." This guidance described a two-year process designed to create support for installation sustainability planning, including a series of four workshops that are recommended to develop an installation sustainability plan. In some cases, the workshops had been combined to help speed the process and save travel costs. The target audiences were installation staff members who are interested in getting a sustainability program started at their installation. This guidance used the Army's strategic planning ABCDEF model and applies it to sustainability. (See Figure 2.1.)

---

[7] Fort Carson, 2002.

Figure 2.1
ISP Workshop and Implementation Process

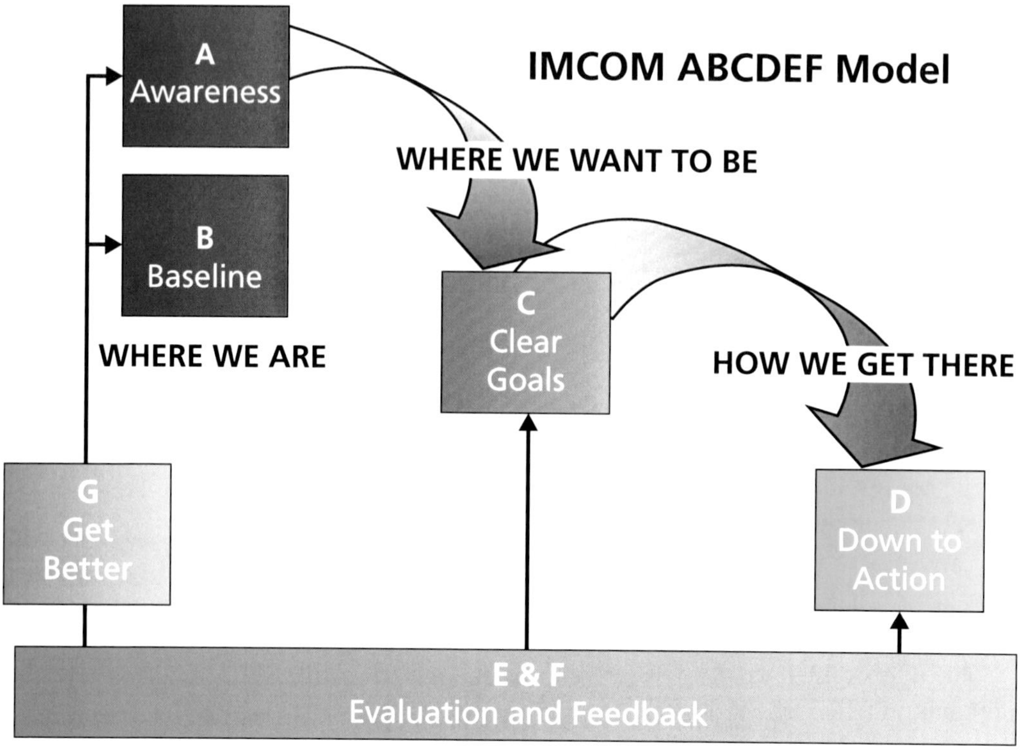

RAND *MG837-2.1*

During the "pre-planning" phase, prior to the first workshop, the guide recommended that installation staff members learn more about the principles of sustainability, recommending a set of references that includes *Dancing with the Tiger: Learning Sustainability Step by Natural Step,*[8] *Natural Capitalism: Creating the Next Industrial Revolution,*[9] and *Believing Cassandra: An Optimist Looks at a Pessimist's World.*[10] Staff are also urged to learn what issues are the most important to the installation, community, and the region. This information should then be used to create a briefing for the garrison commander to secure his or her support for the ISP effort, including scheduling the workshop series. Sustainability proponents should also begin briefing stakeholders on sustainability concepts and approaches to engage them in the process and encourage them to attend the workshops.

8 Nattrass and Altomare, June 2002.

9 Hawken, Lovins, et al., 1999.

10 AtKisson, 1999.

Chapter 2 of the IMCOM Southeast Region's guide recommended that sustainability proponents identify key participants in the process. The Sustainability Planning Team, which guides the entire planning process, should include a planning representative (from the PAIO), a mission representative, and an environmental representative. The Core Teams, organized around each of the installation's core business processes (such as training, transportation, infrastructure, and procurement), will do the planning and implementation work. Stakeholders, who include anyone who is impacted by installation operations, and support players (other Army organizations, external consultants) should be invited to attend the workshops to provide ideas and help develop consensus.

The first workshop—A (Awareness)—brings the process-related teams together for the first time for three half-day sessions to learn about sustainability and identify local issues and challenges. (See Figure 2.2 for a photo of the USAG Hawaii workshop A.) First, teams identify 3 to 5 major activities for each of the core processes. For example, the major activities in the procurement process might include identifying requirements, choosing a procurement method, oversight or quality assurance, and managing downstream waste. Next, teams create system maps that show the relationships between business processes, the community, and the environment. In the third step, the teams list the specific mission, community, and environmental impacts associated with each major activity, as well as the root causes of each impact. In the final team meeting, the teams discuss who needs to be involved to reduce the impacts identified; what data are needed to analyze the effects of actions taken to reduce impacts; and whether any team should be split into two or more teams. The teams develop outbriefs describing the results of each step to inform the garrison commander and get his or her guidance. These outbriefs also form the basis for a draft baseline document as a starting point for the next workshop.

At the second workshop—B (Baseline)—each team reviews the results of the previous workshop and fleshes out the remaining sections of its baseline document during three half-day sessions. The baseline document should be 4–6 pages long and provide a concise analysis of current issues and a snapshot of current conditions. In the first team session, teams develop three concise paragraphs summarizing the key impacts of their processes on mission, community, and environment. In the second session, they draft a "challenge statement" describing key issues to be resolved and the desired end state. For example, at USAG Hawaii, the military training team's challenge statement was "Given our community and environmental concerns and fiscal constraints, how will USAG Hawaii ensure realistic training support done to standard for the warfighter?" In the third session, teams identify current sustainability initiatives in place at the installation and the list of required team members to be invited to the next workshop. USAG Hawaii combined workshops A and B and workshops C and D to save time and travel costs.

Figure 2.2
The USAG Hawaii Awareness Workshop

SOURCE: Photo by Beth Lachman.
RAND *MG837-2.2*

During the third workshop—C (Clear Goals)—each team identifies a set of goals to be achieved over a 25-year time horizon. These are frequently idealistic goals, such as "zero waste" or "100 percent sustainable procurement." This workshop is usually the longest (two and a half days) and the largest of the four. It should include a broader range of stakeholders to generate buy-in and support for the goals and demonstrate the Army's commitment to the long-term success of the installation and the community. The workshop begins with an executive session providing the installation perspective and issues, the community perspective, and a keynote address to motivate participants and educate them on sustainability concepts. It can also include a tour of the installation to explain and show its mission and sustainability challenges.

Participants then break up into teams to develop a set of initial goals that address the challenges identified in the previous workshop. Teams present the goals in a plenary session, and all participants vote on which goals should receive the highest priority. The garrison commander, team leaders, and workshop facilitators review the results, select 6–10 final goals, and reassign them to the teams. The teams meet again to refine the one or two final goals assigned to them and to specify metrics, a time-frame for achieving each goal, the proponents who will be responsible for the goal, and a revised list of team members. Each team briefs its goal(s) to the garrison commander in a final plenary session.

The fourth workshop—D (Down to Action)—is used to complete the installation sustainability plan. Each team meets for one half-day facilitated session, and then meets on its own, if needed, to complete the plan for each of its assigned goals. This involves setting mid-range, manageable objectives over a 5- to 8-year planning horizon that move toward the desired goal; establishing metrics that measure progress toward the objectives and goals; developing action plans that specify who does what tasks by when; setting targets for results to be accomplished in the next 1–2 years; and identifying resource requirements (budget and staff time) for each task.

Implementation of the plan requires Evaluation and Feedback (E and F) to measure progress and periodic updating of objectives and action plans. The IMCOM sustainability guide recommends that the objectives laid out in the ISP be integrated into the installation's existing management system (such as the Army Performance Improvement Criteria, or APIC, system) to engage senior leadership in measuring progress and setting priorities. Objectives and action plans should be integrated into functional plans, such as the Real Property Master Plan, the Range Development Plan, the Integrated Natural Resources Management Plan, and annual work plans that specify how budgets will be executed. Sustainability proponents should continue to educate others who live and work on the installation about sustainability concepts and practices through training classes, post newspapers, sustainability newsletters and annual reports, and other forms of outreach. Team members may need periodic refresher training on sustainability concepts, technical training to implement proposed initiatives, or information on new technologies or approaches that could be used.

At times, the workshop process also helps participants to think more about factors that are important for implementation outside individual traditional areas of focus. For example, an important factor in implementing the ISP is obtaining resources from different sources. The installation may need to leverage resources across funding stovepipes to achieve objectives or develop cross-functional business cases that show the potential payoff from proposed investments. In addition, installations need to develop partnerships with surrounding communities, local suppliers, other installations, and government agencies to help them implement sustainable solutions. In the next section we discuss some examples of approaches that installations have taken in implementing sustainability plans.

How ISPs Have Been Implemented

This section provides an overview of how installations that have developed ISPs have been implementing them. It starts by discussing which installations have begun implementing ISPs. Then it describes how installations have gone about the implementation process.

Which Installations Have Been Implementing ISPs

According to ACSIM, by the end of FY06, 13 installations had developed and were implementing some form of an ISP (see Table 2.2). Some of these installations had begun their ISP process as early as 2000, such as Fort Bragg and Fort Lewis, while

Table 2.2
Installation Sustainable Planning Efforts

Installation Name	State	Major Command Associated with the Installation
Installations that have developed an ISP process		
Fort A.P. Hill	Virginia	FORSCOM
Fort Benning	Georgia	TRADOC
Fort Bragg	North Carolina	FORSCOM
Fort Campbell	Kentucky	FORSCOM
Fort Carson	Colorado	FORSCOM
Fort Eustis	Virginia	TRADOC
Fort Hood	Texas	FORSCOM
Fort Jackson	South Carolina	TRADOC
Fort Knox	Tennessee	TRADOC
Fort Lewis	Washington	FORSCOM
Fort Rucker	Alabama	TRADOC
Fort Stewart	Georgia	FORSCOM
Fort Polk	Louisiana	FORSCOM
Installations supported in FY07 to conduct ISP workshops		
Anniston Army Depot	Alabama	AMC
California National Guard[a]	California	ARNG
Fort Detrick	Maryland	MEDCOM
Letterkenny Army Depot	Pennsylvania	AMC
Pennsylvania National Guard	Pennsylvania	ARNG
USAG Hawaii	Hawaii	IMCOM-Pacific
USAG Wiesbaden	Germany	IMCOM-Europe

[a] The California National Guard is trying to implement its ISP for the entire state, as is the Pennsylvania National Guard, rather than just a particular installation or parcel of land.

others, like Fort Benning and Fort Jackson, had only really begun their implementation activities in 2006, as they completed their workshop process. However, it is important to note that many installation sustainability activities were started and have been ongoing as part of other installation programs before they had a sustainability program. For example, many installations were implementing ecosystem management activities, pollution prevention, and energy-efficiency projects in the 1990s that could be considered part of a sustainability program.

In addition, in FY07, ACSIM provided funding and other support for seven installations to start developing and implementing ISPs (see Table 2.2).[11] ACSIM provided the money and staff for the workshop process to develop an ISP, as described above. In part, these installations were selected to broaden sustainability planning outside of FORSCOM by including the Army Materiel Command (AMC), Army National Guard (ARNG), IMCOM-Pacific, IMCOM-Europe, maintenance depots, and U.S. Army Medical Command (MEDCOM). Some of these installations had completed their ISP development workshop series by the end of FY07, such as USAG Hawaii, while others, like Anniston Army Depot, had not yet begun the workshop process.

How Installations Have Been Implementing ISPs

The approaches used by installations to implement their ISPs have varied. Many have continued to use the team approach that was started by the workshop development process. However, the ISP goals, objectives, implementation teams, organizational responsibilities, project focus and activities, and resources invested vary by installation. As an illustration, we present a detailed description of Fort Bragg's implementation process. Fort Bragg is a natural choice since it was the first installation to develop and implement an ISP, though Fort Hood, Fort Carson, and Fort Lewis all began at almost the same time. Then we briefly discuss two other installations, Fort Lewis and Fort Benning, to compare and contrast their approaches and to illustrate how installations have done some things the same way and other things differently. These three installation examples also include some detailed project descriptions to show the range of implementation activities. The more lengthy project descriptions are placed inside text boxes so as not to interrupt the flow of the main text. Most of these boxes will also be referenced in the next chapter when discussing accomplishments.

Fort Bragg's Sustainability Program. Fort Bragg started working on its sustainability process as early as 1999; it became the official "Sustainable Fort Bragg" program in 2001. Sustainable Fort Bragg management staff is located within the Environmental Management Branch of the Environmental Division under DPW.

The program has evolved over the years, which makes it an interesting example because its focus and structure has changed over time. Table 2.3 shows the 10 original

---

[11] Technically, the Pennsylvania National Guard, USAG Hawaii, and USAG Wiesbaden workshops that were held in 2007 were funded with FY06 year-end funds.

Table 2.3
Sustainable Fort Bragg Goals, 2005

Goal Number	Description
1	Reduce amount of water taken from Little River by 70% by 2025, from current withdrawals of 8.5 million gallons/day.
2	All water discharged from Fort Bragg will meet or exceed North Carolina state high quality water (HQW) standard, by 2025.
3	Landfill waste to be aggressively reduced toward zero by 2025.
4	Meet minimum Platinum standard for all construction by 2020 program, and renovate 25% of all existing structures to at least a Bronze standard by 2020.
5	Adopt compatible land use laws/regulations with local communities by 2005.
6	Reduce energy use in accordance with Executive Order 13123.[a] Specifically, reduce energy use by 30% by 2005 and by 35% by 2010.
7	Develop effective regional commuting options by 2015.
8	Reduce the amount of gasoline and diesel used in nontactical vehicles by 70% by 2015 and by 99% by 2025.
9	Develop an integrated environmental education program for Fort Bragg, its surrounding communities, and interested parties.
10	Work toward 100% Environmentally Preferred Purchasing by 2025 for all purchases, including Government Purchase Cards, contracts, and military requisitions.
11	Create and enhance sustainable training areas that ensure military readiness.

[a] The President, "Executive Order 13123 of June 3, 1999: Greening the Government Through Efficient Energy Management," *Federal Register*, Vol. 64, No. 109, June 8, 1999.

goals that Sustainable Fort Bragg had developed, plus an 11th goal that was redefined in 2005. Goal 11 was initially added to address natural and cultural resources in 2003, after another sustainability workshop was held and identified this need. In December 2003, this goal was stated as "Implement a scientifically-based conservation program for natural and cultural resources compatible with military training and readiness."[12] In 2005, it was changed to "Create and enhance sustainable training areas that ensure military readiness," and to address biodiversity, land use, air space, and water quality concerns related to training lands.[13]

During the first few years, Fort Bragg had 9 to 10 teams that focused on these goals. For example, in 2004 and 2005 the teams included:

- Energy
- Land Use
- Materials
- Sustainable Design

---

12 Fort Bragg, December 2003.

13 Fort Bragg, June 2005.

- Training Lands
- Transportation
- Waste Reduction
- Water Quality
- Water Supply[14]

In late 2003, there was also a natural and cultural resources team. In 2006, Sustainable Fort Bragg reorganized again. At this time, it integrated its sustainability activities into the installation strategic plan. A workshop was held to bring the ISP, strategic planning staff, and other relevant staff together to merge the two plans. The sustainability activities and teams continued during the reorganization, but the goals and teams were slightly reorganized based on the new combined plan as described below.

Sustainable Fort Bragg staff knew they had omitted some issues from the original focus of their activities and the initial 11 sustainability goals. For example, they had left out community/quality of life issues and had not included the Directorate of Information Management (DOIM) in sustainability planning. They have kept their original goals, but have expanded the program to try to focus on more mission and community aspects. A key part of this evolution was the integration of sustainability into the installation strategic plan.[15]

In Fort Bragg's installation strategic plan, there are five overarching installation goals, with the first being focused on a "sustainable community." Garrison goal 1 is defined as: "Fort Bragg—a sustainable community meeting the needs of the Soldier today, tomorrow, and forever." The other garrison goals focus on managing change, customers, financial resources, and leadership.

Figure 2.3 shows the new mission, vision, values, and garrison goals from the June 2006 installation strategic plan. There is a committee, called a garrison goal team, for each of these garrison goals. Each garrison goal team has 12 people on it, representing various installation organizations. Each team also has a Champion and Team Leader. The Champion for the sustainable community goal is the director of DPTM, and the team leader is from DPW.

The sustainability program has six objectives and teams under garrison goal 1, which focuses on creating a sustainable community. The old Sustainable Fort Bragg goals are now subsumed under these new objectives and teams. The "sustainable communities objective teams," as they are called, have subfocus areas, some of which have subteams (see Table 2.4). By summer 2007, Fort Bragg staff were still working on how to organize and define the specific focus areas for the people team. The table shows the proposed focus areas.

---

[14] Fort Bragg, December 2004.

[15] Fort Bragg, June 2006.

Figure 2.3
Fort Bragg Installation Strategic Plan Mission, Vision, and Goals

U.S. ARMY BRAGG

FORT BRAGG

MISSION

Provide the people, infrastructure, and services to train, sustain, mobilize, and rapidly deploy America's forces while enhancing the environment, security, and well-being of the greater Fort Bragg community.

VISION

America's premier military community of enduring excellence emulating

THE SPIRIT OF THE FORT BRAGG SOLDIER

in the quality of our work,
stewardship of our resources,
and commitment of our people.

VALUES

Loyalty: Bear true faith and allegiance to the U.S. Constitution, the Army, and the Fort Bragg community team.

Duty: Fulfill your obligations.

Respect: Treat people as they should be treated.

Selfless Service: Put the welfare of the nation, the Army, and your subordinates before your own.

Honor: Live up to all the Army values.

Integrity: Do what's right, legally and morally. Be responsible for those resources given to you in public trust; it is your duty to improve them for generations to come.

Personal Courage: Face fear, danger, or adversity (physical or moral).

GOALS

1. Fort Bragg—a sustainable community meeting the needs of the Soldier today, tomorrow, and forever.
2. Forecast and manage the effects of change in mission requirements in a systematic manner so that we can marshal our resources to prepare instead of react.
3. Satisfy our customers with a full spectrum of seamless services—always responsive and exceeding their expectations in order to:
    1. Delight the customers.
    2. Be the provider of choice.
4. Invest in the future of our Soldiers and community by ensuring effective stewardship of our resources through
    1. An effective, responsive programming and budget prioritization process.
    2. Use of a system that determines the actual cost for providing customer needs to preclude mission failure.
    3. Leveraging other people's money.
    4. An effective methodology to communicate to higher headquarters our customers' requirements and the associated costs.
5. Fort Bragg—where the Army's Leaders of Tomorrow are forged.

SOURCE: Fort Bragg, June 2006.
RAND *MG837-2.3*

There is a full-time sustainability team coordinator who works with each team. The first five teams also have a full-time sustainability planner. All five planners are subject matter experts from the environmental management branch. Each team has a leader who normally comes from another directorate, (i.e., a different functional area). For example, the land use team focuses on training lands and is led by a trainer. The intent is to be as cross-functional as possible. The transportation team also has members from the surrounding community. There are subteams for the materials/commodities and infrastructure/utilities teams.

All the sustainability teams are tied back to the compliance program through their sustainability planners. Namely, the compliance program pays for the sustainability planners, so they need to be sure that their function and activities help support meeting compliance goals. For example, the transportation planner is part of the

Table 2.4
Fort Bragg Sustainable Communities Objective Teams and Subfocus Areas

Sustainable Communities Objective Team Title	Objective Title	Objective Description	Subfocus Areas
Land use	Land use	Create and enhance sustainable training and urban areas to ensure military readiness and promote compatible growth of the surrounding communities.	Cantonment Training lands Boundaries
Facilities	Sustainable facilities	Become the model sustainable military community of the world by using sustainable principles throughout the life cycle of all facilities and supporting infrastructure (FSI).	New construction Existing buildings Other facilities
Transportation	Sustainable ground transportation	Build a sustainable world-class ground transportation network providing seamless transition between multiple modes of travel while reducing harmful emissions.	Fuel Infrastructure Mass transit
Materials/ commodities	Sustainable materials/ commodities	Achieve zero waste through acquisition and management of materials and commodities which throughout their life cycle creates no additional waste nor requires resources for disposal.	Purchasing Disposal
Infrastructure/ utilities	Reliable, secure utilities	Supply reliable utility services and infrastructure with no negative impact while aggressively reducing over demand. Utilities include energy, water, and information technology.	Water Energy Communications
People	Create a sustainable culture	Create a culture which fosters sustainable lifestyle to enhance the quality of life of the Fort Bragg community. This encompasses the social, mental, physical and spiritual well-being of its members.	Soldiers Family members Garrison staff Community at large

air-quality program and makes sure that projects help address air-quality concerns. However, Fort Bragg is also strategically using its environmental compliance funds to move beyond compliance with sustainability. The sustainability projects themselves, based on their focus, are funded by the Sustainable Range Management (SRM), compliance, or pollution prevention (P2) program. Given the funding sources, most projects have an environmental or training range component to them.

Fort Bragg has a wide range of projects to address its ISP goals and objectives. We briefly provide an overview of some of them here. This discussion is not meant to be comprehensive, but rather to illustrate the range of projects being implemented. The projects are organized by the five team areas. However, it is important to note that projects may fall under and support more than one team area.

Land Use. The land use team has focused on training lands and cantonment areas. The post has started a number of projects to improve the efficiency and effectiveness of training ranges and to reduce their environmental impact. Fort Bragg built an urban training village using painted shipping containers, which saved almost five years of time and an estimated $220,000 compared to standard construction (see Box 2.1).[16] Having this training facility on line quickly was important for warfighting training needs.

Fort Bragg staff are also using recycled plastic for targets, and have converted all of the vehicles in the Range Control vehicle fleet to B20 bio-diesel fuel. They clean and reuse shop rags, and use nonpotable water for hydroseeding in erosion control projects. Another project that helped improve range maintenance was using foxhole covers made from recycled plastics (see Box 2.2).

Fort Bragg is strategically planning cantonment areas to be more sustainable by incorporating sustainable design principles into master planning and facility construction and maintenance. The installation master plan and installation design guide (IDG) include sustainability principles and requirements. The result is that in-site planning and design, building construction, and landscaping activities include sustainability projects. For example, Fort Bragg is building brigade complexes like college campuses to provide more environmentally and community friendly facilities (see Box 2.3).

Facilities. Implementing Leadership in Energy and Environmental Design (LEED) into new construction and existing buildings is a major sustainability effort. The LEED Green Building Rating System is a nationally accepted standard for the design, construction, and operation of high-performance green buildings.[17] Fort Bragg has a "Portfolio" pilot project to implement LEED in 45 existing buildings that were built after 1990. The installation commits to doing certain things throughout these facilities and get LEED credit. This will enable Fort Bragg to build up a portfolio so that they do not have to submit each building for registration individually.

Completed in 2004, the Golden Knights facility became the first "green" building on Fort Bragg. In addition to internal energy saving and other environmentally efficient features, this building includes a below-ground concrete vault that collects storm water, which is then used for irrigation, protecting the nearby lake from storm water runoff and decreasing the use of potable water for irrigation purposes.[18]

---

[16] The time savings is due to the normal timeline required to plan, fund, and build a new training range using military construction (MILCON) funding. It is important to note that the savings cited here were not based on a full life-cycle analysis and did not include an analysis of staff labor costs for the construction of the training village nor an assessment of the full environmental impacts.

[17] For more information about LEED, see Appendix A and the U.S. Green Building Council, "Leadership in Energy and Environmental Design," http://www.usgbc.org/DisplayPage.aspx?CategoryID=19 undated.

[18] The facility was built to a "Gold" standard using the Sustainable Project Rating Tool (SPiRiT), an Army standard that was used before LEED became the official Army standard. See Appendix A for more information about both standards and the Army's evolving green building policies.

Box 2.1
Fort Bragg's Urban Training Village

SOURCE: Photo by Beth Lachman.

Fort Bragg built an urban training village from recycled sealand shipping containers and leftover paint. Fifty containers were stacked and welded together. Range staff acquired these old containers from transportation staff on post at no cost. They mixed 100 gallons of recycled paint to create a beige color, instead of purchasing new paint. This saved them about $1,000. They used 3,500 tons of crushed concrete (for gravel roads, etc.) which saved an estimated $31,500. They also acquired culverts and pipe from the landfill. The project took 90 days instead of the five years it normally takes for the process. Fort Bragg estimated the total cost savings compared to standard construction at about $220,000. However, this estimate was not based on a full life-cycle cost analysis and did not consider any staff labor costs nor the full environmental impacts.

Other facility-related sustainability projects include modifications that help buildings evolve toward LEED "Platinum" standards, such as using recycled carpet squares, installing energy-efficient windows, installing a green roof, and using interior materials, such as paint, that are less polluting and are not airborne irritants.

Box 2.2
Recycled Plastic Foxhole Covers at Fort Bragg

SOURCE: Photo by Beth Lachman.

Across Fort Bragg about 1,000 foxhole covers are needed to cover range training foxholes. Fort Bragg used to use plywood covers, which had to be replaced every two years and painted every year. Cost of replacement was high, and the used plywood went to the landfill. The director of range control got the idea of using recycled plastics instead. He contacted companies that made plastic tabletops about making a plastic foxhole cover and found one that met the specifications. Fort Bragg now purchases plastic foxhole covers made from recycled materials, and the company will take them back to recycle again. They should last nearly indefinitely and do not require painting. Fort Bragg started using them in 2003. Range staff estimate they will save maintenance and other costs totaling $400,000 over 10 years.

Transportation. Fort Bragg is trying to address transportation concerns by using more alternative fuels in nontactical vehicles, developing transportation options and facilities so that people have alternatives to driving a motor vehicle, and developing a comprehensive regional transportation plan. To decrease on-post automobile dependency, Fort Bragg has been incorporating both pedestrian and bicycle facility

Box 2.3
Sustainable Design: Fort Bragg's 3rd BCTC Campus

SOURCE: Photo by Beth Lachman.

Fort Bragg built its 3rd Battle Command Training Center (BCTC) like a college campus. This design is in keeping with Beaux-Arts design principles. The buildings are arranged around a green open space, important buildings are on the axis, and symmetry and balance is in the plan. The brigade headquarters is at one end and the dinning hall at the other. The green space includes a parade field. Parking areas are on the outside and people walk from place to place. This 3rd BCTC campus has four-story barracks surrounding the green space, which provide 1,376 barrack spaces. Some long-leaf pine trees were also saved in the green space. The resulting complex provides visually appealing, pedestrian friendly, and environmentally and community friendly living spaces.

requirements into all construction projects on the installation. These requirements were included in the installation design guide. In addition, to support the current mobilization mission and offer alternatives to driving, Fort Bragg developed and implemented an intra-installation shuttle bus program. The shuttle takes passengers around the installation and connects with the Fayetteville city bus (see Figure 2.4). To address

Figure 2.4
Soldiers from 82nd Replacement Detachment Utilize the Fort Bragg Shuttle Bus Service

SOURCE: Photo courtesy of Fort Bragg.
RAND *MG837-2.4*

regional transportation issues, Fort Bragg staff is working with local, regional, and state transportation and air-quality agencies on mutual public transit and air-quality concerns.

Materials/Commodities. Fort Bragg has a number of activities to address more sustainable purchasing and disposal of materials and commodities, some of which have already been discussed, such as carpet recycling. Other examples include an active recycling and reuse program that reduces solid and hazardous waste, building deconstruction, and green purchasing projects. Fort Bragg has a "Green Procurement Program" (GPP) under which the Assistant Manager of the Fort Bragg Self-Service Supply Center (SSSC) identifies and regularly stocks environmentally preferable products. A recycled label is placed on the shelves to point out "green" products. The SSSC stocks materials with recycled content including printer paper, notepads, toner cartridges, envelopes, and bond paper.

Infrastructure/Utilities. Infrastructure and utilities projects have focused on water, energy, and communications. In addition to the energy-efficiency projects mentioned under facilities and transportation, Fort Bragg is reducing overall energy use, increasing the percentage of renewable energy use, and creating incentives for users to conserve. For example, they are implementing a load management program for natural gas and electric power and installing solar- and wind-powered exterior lights in place of traditional electric lights.

Water projects have included conservation projects, a water reuse plan at the wastewater treatment plant, barrels to collect rainwater runoff from buildings to reduce erosion and stormwater runoff problems, and low-impact development (LID) projects. For example, Figure 2.5 shows a LID project that channels water running off a parking lot through resilient native plants to an underground basin. This process helps slow and absorb water runoff and improve water quality.

Figure 2.5
Fort Bragg Low-Impact Development Project Near Parking Lot

SOURCE: Photo by Beth Lachman.
RAND *MG837-2.5*

Fort Lewis's Sustainability Program. Fort Lewis's sustainability program began in 2000 within the environmental program. Fort Lewis hosted an Installation Sustainability Workshop in February 2002. The purpose of the workshop was to bring together stakeholders from environmental regulatory agencies, the Army, the community, and the installation to build a consensus on Fort Lewis's 25-year sustainability goals. Fort Lewis developed a set of 12 goals and created five teams to develop and begin implementing projects that work toward those goals. Table 2.5 shows Fort Lewis's goals and teams. Fort Lewis's sustainability program is focused heavily on environmental areas, because it is funded out of the installation's environmental program.

As shown by the table, Fort Lewis's sustainability program has some similarities and differences from Fort Bragg's program. Fort Lewis's teams are organized around the traditional environmental areas of water resources, air quality, and energy/infrastructure, whereas Fort Bragg has placed water issues under infrastructure/utilities and

Table 2.5
Fort Lewis Sustainability Teams and Goals in 2007

Sustainability Teams	Sustainability Goals
Air quality	1. Reduce traffic congestion and air emissions by 85% by 2025.
	2. Reduce air pollutants from training without a reduction in training activity.
	3. Reduce stationary source air emissions by 85% by 2025.
Energy/infrastructure	4. Sustain all activities on post using renewable energy sources and generate all electricity on post by 2025.
	5. All facilities adhere to the LEED Platinum standard for sustainable facilities by 2025.
Products and material management	6. Cycle all material use to achieve zero net waste by 2025.
Sustainable training lands	7. Attain healthy, resilient Fort Lewis and regional lands that support training, ecosystem, cultural and economic values by 2025.
	8. Recover all listed and candidate federal species in South Puget Sound Region.
Water resources	9. Zero discharge of wastewaters to Puget Sound by 2025.
	10. Reduce Fort Lewis potable water consumption by 75% by 2025.
	11. Fort Lewis contributes no pollutants to groundwater and has remediated all contaminated groundwater by 2025.
	12. Develop an effective regional aquifer and watershed management program by 2012; completed.

NOTE: These were Fort Lewis's teams and goals in FY07. Since then they have been restructured based on a five-year review of the installation's ISP. In 2008, Fort Lewis had eight goals and six teams, where infrastructure had become its own team entitled "sustainable communities." The latest goals and other changes to the program can be seen at https://sustainablefortlewis.army.mil/Goals.asp.

has a transportation team rather than an air-quality team. Fort Bragg also has a "people" team that is focused on addressing quality of life issues, while Fort Lewis does not.

Both installations have goals to achieve zero waste, reduce air emissions, and implement LEED in facilities. However, Fort Bragg has developed more general objectives for its sustainability program (see Table 2.4). Fort Lewis has focused more on ecosystem, cultural, and natural resource management than Fort Bragg, as demonstrated by goals 7 and 8.

When we examine the activities and projects that are being implemented, more similarities and differences stand out. To illustrate this point, we briefly describe some of Fort Lewis' projects in each team area.

Air Quality. Like Fort Bragg, Fort Lewis has focused on some transportation issues related to air quality. Fort Lewis has been working to reduce traffic congestion by implementing post vans. It has also worked to reduce traffic-related air emissions by purchasing and using alternative-fuel/dual-fuel vehicles in the on-installation GSA fleet. Unlike Fort Bragg, reducing stationary source air emissions is a key part of Fort Lewis's sustainability activities. For example, Fort Lewis's air quality team is constantly looking for innovative ways to reduce air emissions, such as converting boilers from using heavy, high-sulfur fuel oil as a backup heating fuel to using light, clean-burning, low-sulfur distillate fuel oil and using more low-volatile organic compounds in painting operations.

Energy/Infrastructure. Like Fort Bragg, Fort Lewis has been implementing LEED and other energy-efficiency projects in new construction and existing buildings. For example, LEED accredited professionals with the Seattle District of the USACE have integrated themselves into the construction process for the Fort Lewis Whole Barracks Renewal (WBR) Program to continuously improve implementation of sustainability features. By implementing LEED, the WBR Program saved 5 percent of energy usage relative to traditional construction in FY04, and the FY06 WBR was projected to achieve more than 30 percent savings. Other energy projects have included: designing and installing a solar wall for a logistics warehouse as a market demonstration project (see Figure 2.6); implementing energy conservation initiatives, including Direct Digital Controls and the use of high-efficiency condensing boilers throughout new construction; and purchasing 10 percent of power from green sources in 2007.[19]

Products and Material Management. Like Fort Bragg, Fort Lewis has increased the reuse and recycling of construction and other materials. For example, over 9,100 tons of waste concrete and asphalt from construction projects were stockpiled and crushed into aggregate replacement products that are reused for parking lot construction and road repairs. This recycling activity results in approximately $340,500 annual savings in disposal fees and the cost of purchasing similar material, as well as a reduction of solid waste generated.[20]

---

[19] By spring 2008, Fort Lewis was purchasing 21 percent of its power from green sources.

[20] Fort Lewis, 2006.

Figure 2.6
Fort Lewis Solar Wall on a Logistics Warehouse

SOURCE: Photo by Beth Lachman.
RAND *MG837-2.6*

Fort Lewis is composting biosolids from the wastewater treatment facility, wood waste, destroyed classified document media, grass clippings, leaves, and horse manure to create a soil amendment landscaping product.

Fort Lewis pioneered a project to purchase more sustainable furnishings for offices on post, called the Sustainable Interiors Showroom (SIS). (See Box 2.4.) Fort Bragg has not implemented a SIS project, but other installations, such as Fort Jackson, have, based on Fort Lewis's experience.

Unlike Fort Bragg, Fort Lewis includes a program to supply native plants for landscaping under its products and material management team. In 2006, the Fort Lewis Fish and Wildlife Program supplied 1,600 native plants to landscape the Sequalitchew Creek EcoPark entrance.[21] However, Fort Bragg is planting native landscape as part of its sustainable design projects under its land use team's cantonment activities.

21 Fort Lewis, 2006.

Box 2.4
Fort Lewis Sustainable Interiors Showroom Facilitates Purchase of Sustainable Furnishings

The Sustainable Interiors Showroom is a sustainable product demonstration site displaying flooring materials, office furniture, paint, and lighting from GSA vendors. Installation purchasing staff can tour the showroom to learn about sustainable products, and may choose to purchase some of these products. Availability of the SIS resulted in the purchase of more than $180,000 in recyclable and/or recycled-content furnishings for several Fort Lewis units and facilities, including the Soldier Readiness Processing Site and Stone Education Center. In addition, use of recyclable carpeting squares as replacement for existing nonrecyclable floor covering is now an accepted business practice for most new projects. At least 260,000 square feet of recyclable carpeting was installed at Fort Lewis in 2005.

Sustainable Training Lands. Fort Lewis is undertaking conservation and ecosystem management and restoration activities to try to prevent threatened and endangered species (T&ES) concerns from limiting training range operations. Unlike Fort Bragg and most other installations' sustainability programs, natural resource and conservation activities are a major component of Fort Lewis's ISP program.[22]

An important part of Fort Lewis's sustainability program for training lands is to consider encroachment effects from nearby urban sprawl and the loss of biodiversity. Fort Lewis is participating in the Army Compatible Use Buffer (ACUB) program to help create a buffer against encroachment. ACUB allows Army installations to use funds to enter into partnership agreements with county, state, or municipal governments, as well as nonprofit organizations, for the partner to purchase tracts of land or easements on lands from willing sellers to establish buffers around installations to maintain current land uses or to protect habitat. These buffers help limit the effects of encroachment and maximize use of land resources in support of the installation's mission. Fort Lewis is using ACUB to help preserve and restore native prairie habitat and species in partnership with The Nature Conservancy (TNC) and the Washington state departments of Natural Resources (WDNR) and Fish and Wildlife (WDFW). TNC, with assistance from WDFW and the U.S. Fish and Wildlife Service, acquired a 127-acre prairie preserve through ACUB. Under the ACUB program, Fort Lewis is working to manage additional prairie land in the southern Puget Lowlands acquired by private land conservation groups; restore native prairie on these lands and other prairies that are already protected; and begin reintroduction of the four listed candidate species: the Mardon skipper, Taylor's checkerspot (see Figure 2.7), the streaked horned lark, and the Mazama pocket gopher.

---

[22] While Fort Bragg has a strong natural resource management program, including efforts to restore long-leaf pine (LLP) habitat, recover red-cockaded woodpeckers (RCWs), and perform ACUB conservation buffering, in 2007 these activities were not being integrated into the ISP process.

Figure 2.7
The Taylor's Checkerspot Butterfly

SOURCE: Photo courtesy Fort Lewis.
RAND *MG837-2.7*

Other natural resource–related sustainability activities to support training include locating and abating Scotch Broom and other invasive species on training lands and growing native grass plants such as Roemer's Fescue and planting them in training areas. Fort Lewis even provided native grass seeds to a Forestry Service nursery.

Water Resources. Like Fort Bragg, Fort Lewis has undertaken a number of projects to address water quantity and quality concerns. However, specific projects are different from Fort Bragg's activities and have included working with local communities on watershed planning, remediating contaminated ground water with on-site electrical heating technology, and installing reclaimed water pipes. For example, reclaimed water pipe has been incorporated into some new whole barracks renewal projects since FY02, allowing the reuse of rainwater for facility nonpotable water needs; this activity is also included in the energy and infrastructure team's LEED standards.

Fort Benning's Sustainability Program. Fort Benning's sustainability program is organized differently from those of Fort Bragg and Fort Lewis, with a very different team structure and set of goals. The post's ISP was developed by the Fort Benning sustainability team members during facilitated workshops and sustainability conferences in 2005 and 2006. Information collected during those workshops and conferences was incorporated into current sustainability goals and objectives, along with initial actions that Fort Benning planned to take toward those goals. Fort Benning's plan is also different in that it includes specific details for reaching each objective, including: specific performance measures, who owns the objective (i.e., has responsibility for its implementation), specific targets, and detailed action plans.

Fort Benning's sustainability program is divided into four teams: regional interaction, military training, installation management, and power projection. Table 2.6 shows the goals by team area, as presented by the installation sustainability/garrison strategic plan. From the start, Fort Benning integrated its sustainability plan into its garrison strategic plan and had a much broader focus than other ISPs. The Fort Benning sustainability coordinator is a strategic planner and a member of the garrison commander's staff. There is both garrison and environmental funding for sustainability activities.

Table 2.6
Fort Benning's Sustainability Teams and Goals

Team	Goal
Regional interaction	A Chattahoochee Valley community that sustains the Fort Benning mission, enhances quality of life, and protects and restores the environment will better serve the overall objectives.
Military training	Increase training space (air, land, water, and bandwidth) by 50%.
	Fort Benning becomes the Army Live-Virtual-Constructive (LVC) Center of Excellence for Joint and Combined Arms Operations.
	Establish world's most innovative Maneuver Center of Excellence.
Installation management	Achieve procurement of 100% sustainable goods and services by establishing an effective procurement network that minimizes life-cycle costs, maximizes acquisition options, reduces delays, and establishes systemwide accountability and ownership.
	Fort Benning leads DoD in the provision of Soldier and Family Support Services.
	Capture full economic potential for energy efficiency through the use of innovative and sustainable approaches to energy acquisition, management, and consumption.
	Implement sustainable water acquisition, use, and management practices that support the mission of Fort Benning.
	Facilities at Fort Benning meet sustainability objectives.
Power projection	Increase deployment capacity and decrease deployment time for brigade elements by 2030 to 25% of FORSCOM standards.
	Eliminate frustrated cargo and decrease deployment time through reduction and improved management of HAZMAT on the installation and during deployment.

Fort Benning's goals and activities focus on a range of mission areas in addition to improving training lands, such as deployment. Fort Benning also has much more emphasis on quality of life issues in its ISP implementation, in comparison to the other implementations described, as demonstrated by the goal: "Fort Benning leads DoD in the provision of Soldier and Family Support Services."

Fort Benning's implementation activities have some similarities to those at Forts Bragg and Lewis, but also many differences. To illustrate these differences and similarities, we provide a brief overview of some of Fort Benning's sustainability activities. Since Fort Benning's ISP program started relatively recently, many of these activities have not yet started, so we only discuss their objectives and plans here. Again, we organize this discussion around the installation's team areas.

Regional Interaction. Fort Benning's regional interaction activities are focused on concerns about urban growth, surrounding land use, quality of life concerns, and relationships with the surrounding community. The team's focus is best summed up by the regional interaction challenge statement from Fort Benning's ISP:

> Sustained long-term growth and periodic variations in growth patterns will result in changing land use and potential urban sprawl that threaten the use of installation training lands and negatively impact the quality of life in the region. These factors have led to increased demand for natural resources and utilities, which have the potential to degrade water, air quality, and green space/habitat. In addition, Fort Benning is transitioning to a greater reliance on the community for basic services like education, health care, public services, etc.
>
> How does Fort Benning strengthen partnerships with the regional community in order to maximize and sustain its training and deployment missions, quality of life for soldiers, family and neighbors, and protect/enhance the environment?[23]

Sustainability activities include ACUB, noise abatement procedures to minimize community noise complaints, and regional outreach and education on sustainability. For example, Fort Benning has developed an ACUB proposal to acquire land or non-development easements to create a three-mile buffer space between military activities and the surrounding community.[24] This team is also focused on access to sustainability resources, which include funding, skills, commodity and service providers, and developing opportunities to create markets for local businesses.

Fort Benning has some projects managed by its regional interaction team to address transportation and air-quality concerns that are similar to those of Forts Bragg and Lewis. Such projects and activities include purchasing about 300 alternative fuel

---

[23] Fort Benning, 15 February 2007, p. 40.

[24] Fort Benning, 15 February 2007, p. 42.

vehicles (AFVs) for part of the GSA-provided vehicle fleet, trying to encourage carpooling, and replacing 80 gas-powered golf carts with electric golf carts.

Military Training. Fort Benning's military training sustainability activities are focused on developing and maintaining state-of-the-art training facilities and lands. Unlike other installations, Fort Benning places an emphasis on developing the training facilities themselves using improved technologies, such as creating a Live-Virtual-Constructive strategy and infrastructure that may include linking together hardware, software, buildings, secure networks, range instrumentation, and other emerging technologies. Like Fort Lewis, Fort Benning's training sustainability activities include a focus on natural resources and ecosystem management. One of the team's objectives is "sustainable cultural and natural resources." In 2007, activities to support this objective include the Integrated Natural Resource Management Plan (INRMP) natural resource and ecosystem management efforts to help restore the LLP ecosystem and manage habitat for gopher tortoises (see Figure 2.8), a key species of concern, and the Land

Figure 2.8
Gopher Tortoise on Fort Benning Training Range

SOURCE: Photo by Beth Lachman.
RAND *MG837-2.8*

Rehabilitation and Maintenance (LRAM) and Integrated Training Area Management (ITAM) programs. LRAM and ITAM projects are designed to achieve sustainable use of training lands by implementing a uniform program that inventories and monitors land conditions, determines the carrying capacity of the land to meet training requirements, and provides for land rehabilitation and maintenance measures.

Installation Management. Fort Benning has grouped many sustainability activities under this general category, including energy projects, green buildings, construction and demolition waste projects, storm water and wastewater management, the purchase of environmentally preferable products, hazardous waste reduction, and recycling activities. For instance, Fort Benning is participating in the Residential Communities Initiative (RCI), which transfers ownership, construction, maintenance, and operation of military family housing to large housing contractors through 50-year contracts. Fort Benning plans to have the housing privatization partner incorporate sustainable design and development principles and practices into the design, construction, and operation of family housing. Natural gas, water, and electricity will be metered at each housing unit, which will help to promote energy and water conservation. Similar to Forts Bragg and Lewis, Fort Benning staff are developing a Green Procurement Program, grinding concrete and masonry rubble for aggregate, implementing LEED in new construction, and actively recycling solvents, used oils, and construction and demolition (C&D) waste.

However, unlike these two installations, Fort Benning also includes a major focus on soldier and family support. One of the objectives is to make soldier and family support facilities the "best in DoD." Planned activities to support this objective include: building a new hospital, expanding the Commissary, Morale, Welfare, and Recreation (MWR), and lodging facilities (hotel, gym/fitness, and recreation centers) and providing a single "easily identifiable and accessible facility that houses all soldier and family support service organizations for soldiers, family members, retirees, and civilians (e.g., centralized in/out processing)."[25] Fort Benning's ISP also includes an objective to meet the education needs of family members with either DoD schools on the installation or through partnerships with local communities.

Power Projection. Unlike the other installations, Fort Benning has a sustainability program that focuses on its mobilization and deployment missions. Fort Benning is working to ensure that its infrastructure, services, and lands are adequately managed to concurrently deploy multiple units by air, rail, and ship, and to support all needs of mobilization/deployment, unit sustainment, and Program of Instruction (POI) training requirements. In 2007, sustainability efforts include improving the airfield, port, and railway systems. For example, Fort Benning is exploring means to ensure that rail capacity can best serve the deployment mission. Fort Benning staff also are working to "optimize installation loading and material handling systems to support simultaneous

---

[25] Fort Benning, 15 February 2007, p. 83.

air and land deployments (e.g., on- and off-road loaders, equipment, portable ramps, rollers, docks, equipment storage, and qualified hazmat-handling personnel)."[26]

There are also environmental technology and process improvement activities included under the power projection area, such as concrete grinding, hazardous materials/waste reduction, solvent recycling, and fuel-efficiency projects.

Conclusions for the Installation Comparison. By comparing and contrasting the goals and objectives, team organizations, and activities of these three installations, it is clear that there are many similarities and differences in the approaches and activities of these ISP efforts. To illustrate this point with a summary table, see Table 2.7, which compares the sustainability team organizations for these three installations.

It is important to acknowledge that each installation has been focusing on what will work best given the unique culture, resources, and environment of that post, surrounding community, and region. Variance is expected because of differences in local and regional conditions, including varying environmental concerns, missions, funding, installation infrastructure and facilities, community relations, and internal installation organizations and processes. In addition, variability in ISPs is expected, according to the management literature, because the ISP process is still in the pilot and experimental stage before becoming a more widespread and developed practice.

Table 2.7
Comparing Sustainability Teams

	Team Name		
Issues Covered	**Fort Benning Sustainability Team**	**Fort Bragg Sustainable Communities Objective Team**	**Fort Lewis Sustainability Team**
Land use		X	
Training lands			X
Military training	X		
Power projection	X		
Installation management	X		
Facilities		X	
Infrastructure/utilities/energy		X	X
Transportation		X	
Air quality			X
Water resources/quality			X
Materials/commodities/products		X	X
People		X	
Regional/community interaction	X		

[26] Fort Benning, 15 February 2007, p. 106.

CHAPTER THREE

Installation Progress in Implementing Sustainability

To understand the advantages and disadvantages of the pre-FY08 ISP development and implementation processes, it is important to more broadly examine the progress that installations have made in implementing ISPs. We assessed initial progress in implementing ISPs, based on installation visits, on reviews of installation sustainability documentation and sustainability and other literature, and on phone interviews with installation sustainability and other staff. We assessed whether installations were implementing projects and making some initial progress toward their own stated goals and the Army's broader sustainability goals. We found that many installations are making progress, but challenges remain. This chapter describes the progress, while the next discusses challenges and needs for improved implementation based on the progress and challenges we observed.

Installations Have Made Progress in Implementing Sustainability

Installations have accomplished a great deal in implementing their ISPs and sustainability activities, given limited resources and guidance. Below we illustrate some of these accomplishments. This is not a comprehensive assessment of accomplishments, but an initial assessment of progress and an illustration of the range and scope of accomplishments based on interviews with installation and other Army staff and a literature review. A comprehensive assessment would involve an in-depth analysis of the full range of the activities' effects on sustainability, accounting for all of the activities' costs and benefits to the installation, environment, and the broader community. It would also involve assessing whether appropriate metrics were set and tracked for each goal at each installation and for the broader Army sustainability goals.[1]

Based on installation and Army sustainability goals and the broader sustainability literature, we also assessed and discussed some sustainability areas where there have been fewer achievements and some of the reasons why. This discussion is important to identify the gaps that could be addressed by HQDA guidance and other assistance to

[1] Given our resources and focus, such a comprehensive assessment was outside the scope of our study.

installations implementing ISPs. Chapter Four will supply more detail on the barriers that can slow down or prevent ISP implementation.

Installations with the longest history and the most support have demonstrated the most success, including: Fort Bragg, Fort Carson, Fort Hood, and Fort Lewis. However, other installations have had some accomplishments as well, as we will discuss.

The accomplishments are organized into the following sustainability areas:

- Mission
- Facilities/buildings
- Transportation
- Infrastructure
- Products and materials management
- Natural resources management
- Community/regional interaction
- Quality of life issues

These categories were chosen based on our review of ISP development and implementation structures, projects, and other activities that installations are implementing, as well as the sustainability literature. They seemed most appropriate given the 2007 ISP activities. Others might be appropriate as well, such as breaking mission or quality of life into multiple categories, as Fort Benning did with its two mission-related teams, military training and power projection. Another approach could be to focus more on functional areas rather than the general categories of mission and quality of life to better examine integration across mission, community, and environmental areas. However, the list above includes the most commonly used categories, given installations' 2007 implementation activities.

In each category, we examined the number of installation sustainability activities in each category, the extent of the coverage and range of activities in the category to help identify gaps, and what had been accomplished so far based on installation and Army sustainability goals and the broader sustainability literature.

Mission

Since most of the installations that have been implementing ISPs so far have a training function, this has been the main area of mission accomplishments. Installations such as Forts Bragg, Carson, Hood, and Lewis have had a number of accomplishments in training range development, management, and maintenance. As discussed in the last chapter, Fort Bragg's construction of an urban training village (Box 2.1) is a good example of a range development accomplishment.

Many range sustainability activities have successfully helped address training range erosion and other maintenance issues. For example, Fort Hood has built hill access trails to facilitate tank access and reduce erosion and used "gully plugs" (see Box 3.1) to create training vehicle bridges, help improve water quality, and address erosion problems.

Box 3.1
Fort Hood Gully Plugs Help Address Erosion Problems

SOURCE: Photo by Beth Lachman.

Fort Hood has had significant problems with erosion and water runoff problems from gullies that form in maneuver areas. A gully is a network that causes erosion on the landscape. Vehicles cannot cross gullies if they are deeper than two feet for wheeled vehicles or three feet for tanks. The vehicle will flip over or get stuck in the gully mud. Fort Hood developed an innovative approach to address problems with gullies. It created a "gully plug" by shaping the edges of the gully and placing large stones in the gully hole. These rocks come from a local quarry, and make a natural bridge that vehicles can cross, i.e., a tank trail across the gully. The gully plug also slows water movement and causes sediment to fall through the rocks, which helps with water quality concerns. Nature then repairs the system over time. The silt in the gully fills up behind the gully plug and soil develops between the rocks. Over a period of a few years, grass and other vegetation grows on the plugs and the land heals itself. The plugs also provide better drainage because of the rocks under the new soil. Fort Hood built about 400 gully plugs in 2006. The process is faster and less expensive than the previous method of gully repair.

Another erosion-control project that saves time on maintenance, reduces erosion, and helps improve water quality is a green firing range built by Fort Bragg. Normally, firing ranges have tall berms with steep slopes so that no grass grows on the sides, resulting in erosion problems that cause constant maintenance issues. Fort Bragg range staff reduced the berm height and regraded it with a more gradual slope so that grass grows on it, reducing erosion problems and maintenance work. They put in a drainage system with plastic pipes, similar to a septic system. They also have a catch basin in back that creates a pond. Staff biologists are studying this retention pond because of an endangered frog species in the area.[2]

Reducing the environmental impact of the maintenance of training facilities is also a common area of accomplishments, such as Fort Bragg's green firing range and recycled plastic foxhole covers (Box 2.2). A more common example is the recycling of training range materials. For instance, Fort A.P. Hill and a couple of other installations have successfully recycled scrap metal targets from their ranges. As of September 2007, Fort A.P. Hill staff had recycled 7,300 tons of scrap metal and installed 16 "environmentally friendly green targets." Fort A.P. Hill staff estimate that this effort has saved $563,000 and landfill space.[3]

An example of an accomplishment that reduces both environmental and soldier health effects from training operations is Fort Hood's use of recycled tires to create a platform for a tank firing range to reduce dust, which improves air quality. The tank main gun barrel is above the rubber platform as it fires and the rubber absorbs the firing shock wave so no dust is generated. (See Figure 3.1.)

Installations also have accomplishments related to mission vehicle maintenance and industrial functions. However, most of these are in traditional environmental areas, such as pollution prevention, recycling, and reduction of solid and hazardous wastes related to mission operations. For instance, Fort Hood has an extensive program to reuse and recycle solid and hazardous materials, including training range materials. Facilities located on the ranges collect commonly used items, such as oil, antifreeze, and paints (see Figure 3.2), for reuse.

Installations are also working to address encroachment threats that may place limits on testing and training operations. Some installations, such as Forts Benning and Carson, include these activities in their sustainability program, while others do not. Some installations, including Forts Carson, Lewis, and Stewart, have had some initial success through the ACUB program. For example, Fort Carson has permanent conservation easements on 12,000 acres, and Fort Stewart has protected over 420 acres of open space with conservation easements on private lands near the installations.[4]

---

2 Information provided by Fort Bragg range control staff.

3 Information provided by Fort A.P. Hill staff.

4 For more information about these two installations' conservation buffering accomplishments, see Appendixes C and D in Lachman et al., 2007.

Figure 3.1
Recycled Tire Platform at Fort Hood Tank Firing Range

SOURCE: Photo by Beth Lachman.
RAND *MG837-3.1*

Very few installations have focused on other mission functions, such as deployment. An exception is Fort Benning, which, as discussed earlier, includes power projection as part of its sustainability program. Fort Benning's accomplishments in this area include expanding its airfield to accommodate all commercial jets as well as Air Force C-5s and C-17s, increasing loading facilities and capacity, and creating a new passenger processing terminal.

Facilities/Buildings

Many installations have incorporated more sustainable practices into the design and construction of individual buildings and facilities. The most common approach is adopting LEED standards and other types of green building activities. One example is the U.S. Army Medical Department Activity (MEDDAC) barracks at Fort Carson. The barracks is a SPiRiT Gold rated facility (the Army's green building standard before it adopted LEED). It is located next to Evans Army Community Hospital, so soldiers

Figure 3.2
Used Oil Collection Facility at Fort Hood Training Range

SOURCE: Photo by Beth Lachman.
RAND *MG837-3.2*

can walk to work. Additional sustainable features include a building orientation to maximize solar gain, low-flow water fixtures, and drip irrigation of a surrounding xeriscape landscape.[5]

Other sustainable building accomplishments include the Golden Knights facility at Fort Bragg, LEED implementation in Fort Lewis's Whole Barracks Renewal Program, and two buildings made of agriboard at Fort Hood. Agriboard is structural insulated panels made from compressed wheat or rice straw that are twice as energy efficient as wood and more fire and pest resistant than wood. On another building, Fort Hood installed a sustainable roofing product with high solar reflectance and with minimal waste compared to the traditional process of applying tar and asphalt shingles to roofs. It keeps the building cooler, saves energy, and had a cost avoidance of

5 See Fort Carson, 2006, p. 7, and Galentine, undated (c).

$400,000.[6] This example illustrates the practice of installing a more sustainable feature within an existing building; other examples related to energy, water, and green procurement are discussed below in the subsections on infrastructure and products and materials management.

Other facilities are also being designed to be more sustainable, including golf courses. For example, in November 2005, Fort Bragg's Ryder golf course became a "Certified Audubon Cooperative Sanctuary." To be certified, the course and its management practices had to become more sustainable, which was done through enhancement of valuable natural areas and wildlife habitats and reduced water and chemical use.

Addressing air quality emissions at stationary facilities is another common type of accomplishment because of installations' strong environmental programs and the need to comply with the Clean Air Act Amendments. For example, Fort Lewis's efforts at converting boilers to cleaner-burning fuel oils and by switching to low-volatile organic compounds in painting operations has contributed to a significant decrease in air emissions.[7]

Only a couple of installations have implemented sustainability more broadly into master planning to design more sustainability communities, not just individual facilities and buildings. One example is Fort Bragg, which incorporated sustainability principles into its 2004 Real Property Master Plan (RPMP). In fact, this plan incorporates Sustainable Fort Bragg into its vision statement: "In order to maintain Fort Bragg's legacy and continue to train troops to standard, long-term sustainable development principles and practices must be incorporated into Fort Bragg's day-to-day operations."[8] Chapter 3 of the plan also focuses on sustainability. This approach is reflected by Fort Bragg's design of more pedestrian and environmentally friendly and community-enhancing barracks complexes, as discussed in Box 2.3. Similarly, Fort Bragg is building junior enlisted homes to resemble historic 1930s bungalows (see Figure 3.3) and has a surrounding landscape plan to promote a more aesthetically appealing and pleasant community. Fort Lewis has also updated its master plan to reflect sustainability goals.

Fort Bragg, Fort Hood, and Fort Lewis have all integrated sustainability into their installation design guides, which provide information to contractors about installation design and development decisions.[9]

Fort Carson has also made some progress incorporating sustainability into its master planning by developing a "Town Center" concept. The "Town Center" conceptual plans show a build-out of the area from the Post Exchange to the Commissary with a main street that connects these two facilities. Retail, housing, office space,

[6] Luciano, July 2007.

[7] Fort Lewis, 2006.

[8] Fort Bragg, February 2004, p. 2-1.

[9] For more information see Fort Bragg, undated.

Figure 3.3
Historic Junior Enlisted Home at Fort Bragg

SOURCE: Photo by Beth Lachman.
RAND *MG837-3.3*

recreational facilities, medical facilities, and other types of support facilities have been considered in the early stages of this planning effort. As a key partner, the Army and Air Force Exchange Service (AAFES) is studying the "Town Center" concept at Fort Carson to determine its economic feasibility.[10] In 2007, this Town Center concept evolved into a "Lifestyle Village" concept, which includes many of the same sustainability aspects.[11] Since it is not implemented yet, it is not a complete accomplishment; however, having a plan is an initial accomplishment.

Transportation

Some installations have focused on transportation issues to provide alternatives to single vehicle ridership, to address air quality and other environmental concerns, and

---

[10] Fort Carson, 2006, p. 7.

[11] Fort Carson, 2007, p. 8. Also see Jackson, May 31, 2007.

to address regional transportation congestion and growth issues. However, accomplishments so far are limited and mostly affect transportation internal to the installation.

Some installations, such as Forts Bragg, Carson, and Lewis, have implemented post shuttle vans or buses. However, due to severe budget constraints and limited ridership, Fort Carson discontinued its post shuttle bus in June 2006.[12] Fort Bragg's and Fort Lewis's efforts have been more successful. By 2006, Fort Lewis had increased its post rideshare program to 19 vans and 190 participants, and the demand exceeded the supply of vans from local transit agencies.[13]

Use of alternative fuel vehicles is the most common transportation-related accomplishment. This practice can potentially help address air quality, climate change, and energy concerns.[14] We give three examples of such accomplishments for Forts Lewis, Carson, and Bragg respectively. Fort Lewis reduced traffic-related air emissions during 2005–2006 by increasing the percentage of alternate-fuel/dual-fuel vehicles in its GSA fleet to 40 percent.[15] Similarly, Fort Carson has increased the use of alternative fuels by investing in the needed infrastructure and requiring the use of E-85 fuel. At Fort Carson, an E-85 fuel pump was installed in August 2006 at the contractor-owned/contractor-operated fuel facility on the post to allow alternative fuel vehicles to use the less polluting E-85 fuel. Previously, E-85 fuel was only available from a remote off-site location. The E-85-capable GSA vehicles on the installation are required to use E-85 fuel rather than regular petroleum-based fuel.[16]

Similarly, Fort Bragg Range Control converted every vehicle in its fleet, about 12–15 vehicles, so they could use B20 diesel fuel. In 2003, Range Control was using a mixture of about 50-50 bio-diesel/regular diesel. However, the bio-diesel price started rising, so they stopped using it as much. Then usage went back up to about 85 percent in early 2007.[17] This example shows how accomplishments can vary from year to year based on external factors, such as the relative prices of bio-diesel and regular diesel fuels.

Addressing traffic congestion, providing alternative modes of transportation (such as bicycles and public transit), and encouraging people to use them is a difficult

---

[12] Fort Carson, 2006, p. 3.

[13] Fort Lewis, 2006.

[14] It is important to note that the full sustainability costs of individual alternative fuels used would need to be assessed before claiming true benefits for climate change and sustainability. For example, corn ethanol production can potentially cause a range of environmental problems, including soil erosion; large water use; water and soil pollution from the fertilizers and pesticides used to grow the corn; large uses of land, potentially resulting in habitat and biodiversity loss; increasing grain and food prices throughout the world; and potentially even increased carbon dioxide emissions from the energy inputs to grow the corn (Schulz, 2007 and Buntrock, 2007).

[15] Fort Lewis, 2006.

[16] Fort Carson, 2006, p. 3.

[17] Information provided by Fort Bragg staff.

sustainability challenge that requires long-term strategic planning and collaboration with surrounding communities on transportation infrastructure and land-use planning. In 2007, only a few installations, such as Forts Bragg and Carson, were trying to address these issues. For instance, Fort Carson's 2006 sustainability report states, "The impact of increasing traffic congestion and transportation needs are growing concerns for Fort Carson and the Pikes Peak Region. The Installation is pursuing long-range strategies to not only address the near-term growth, but also plan for future transportation requirements."[18] However, these installations had only had minimal accomplishments in 2007. Fort Carson had made some progress at improving traffic flow and encouraging alternative modes of transportation. Installation entry gate projects were implemented to accommodate pedestrian and bicycle traffic, and extra approach lanes were added to two gates to decrease fuel wasted by excessive idling and to improve overall traffic flow. However, Fort Carson did not yet have any accomplishments with respect to regional transportation management with the surrounding communities. Fort Bragg has incorporated pedestrian and bicycle facility requirements and considerations into master planning and new buildings, such as the 3rd BCTC campus design (see Box 2.3). Fort Bragg also is participating in a Base Realignment and Closure (BRAC) transportation planning process that is trying to develop regional transportation corridors. However, in 2007, like Fort Carson, Fort Bragg did not yet have any accomplishments in regional transportation planning.

Infrastructure

For the purposes of this discussion, we define infrastructure as Fort Bragg did: energy, water, and communications. Many installations have accomplishments in energy and water sustainability, but almost none have made progress in communications, since most do not appear to include communications infrastructure in their sustainability activities. Here we discuss the range of energy and water accomplishments.

Many installations have been implementing more energy-efficient and sustainable technologies and practices in buildings, lighting systems, and vehicles, and have demonstrated accomplishments in these areas. Many of these projects have been implemented because installations have had strong, long-term programs in energy management and efficiency because of the potential cost savings, as well as Army and other federal energy policies, such as the Energy Policy Act of 2005.[19] For example, combined energy initiatives for FY06 saved Fort Carson approximately $1 million.[20] We illustrate the range of energy project accomplishments (except for vehicles, which were

---

[18] Fort Carson, 2006, p. 3.

[19] For more information on such policies, see the "Army Energy Program" web site at http://army-energy.hqda.pentagon.mil/policies/epa.asp

[20] Fort Carson, 2006, p. 2.

already discussed in the transportation subsection) with examples from five different installations.

Typical energy-efficiency and conservation accomplishments are demonstrated by Fort A.P. Hill, which added remotely controlled thermostats in 57 buildings, increased insulation at 24 barracks buildings, and is replacing older boilers with new energy-saving models. More efficient lighting is being tested at Fort Benning (photovoltaic lighting units) and Fort Hood (solar-powered parking lot lights). Fort Hood achieved significant energy savings by changing lighting practices at Gray Army Airfield. Its runway lights were kept on 24 hours a day, even though the airfield was not used all the time. Engineering division staff installed LED lights and reflectors and radio controllers for power activation. Now pilots can tune into the required frequency to turn the power and lights on. The result was a savings of over $60,000 a year and a 63 percent reduction in electricity consumption.[21]

Several installations have accomplishments in implementing large-scale solar technology projects. Forts Lewis and Carson have installed solar walls (see Figure 2.6). At Fort Carson, a transpired solar collector wall was installed on a large motor pool building to reduce natural gas consumption by preheating intake air for the building's heating system. The wall is estimated to save $25,000 to $35,000 per year in natural gas consumption.[22] USAG Hawaii has implemented a large-scale solar project in family housing through a contractor-financed Residential Communities Initiative project, making it one of the largest solar-powered communities in the world. Photovoltaic panels will provide approximately 30 percent of all family housing energy needs.[23]

A fairly unique energy accomplishment at Fort Carson has been the purchase of green power certificates. Fort Carson is in its third year of a five-year contract to purchase Renewable Energy Certificates (RECs) worth 40,000 megawatt-hours per year from wind and biomass energy through the Western Area Power Administration. In 2006, the REC purchase accounted for 28 percent of Fort Carson's electrical use.[24]

Fort Carson also has had a twelve-acre 2 MW solar array built, which in 2008 was the largest of its kind in the Army. In FY07 construction began on this contractor-owned/contractor-operated 2 MW, ground-mounted solar photovoltaic array to generate electricity for Fort Carson (see Figure 3.4). It is estimated that the array will generate 3,200 megawatt-hours (MWh) of power annually, which equates to approximately 2.3 percent of Fort Carson's energy consumption, or enough to provide power to the equivalent of 540 Fort Carson homes per year.[25]

---

[21] Luciano, July 2007.

[22] Fort Carson, 2006, p. 2.

[23] U.S. Army, undated (c).

[24] Fort Carson, 2006, p. 2.

[25] Fort Carson, undated.

Figure 3.4
The 2-Megawatt Solar Array at Fort Carson

SOURCE: Photo courtesy of Fort Carson.
RAND *MG837-3.4*

Some installations have accomplishments at implementing projects to address water quality and supply concerns. The purposes of these projects include conserving water usage, such as installing waterless urinals at Forts Carson, Benning, and Hood; reducing storm water runoff; and reusing water.

Both Fort Bragg and Fort Hood have had some success at addressing water runoff concerns. Fort Bragg has implemented LID projects to help reduce stormwater runoff (see Figure 2.5). Because of sediment and pollution concerns about installation water runoff into nearby Belton Lake, a drinking water source for the community, Fort Hood has reduced concentrated water flow by about 60 percent and reduced sediment movement by 90 percent using best management practices on its ranges.[26]

Fort Lewis has a range of water resource accomplishments. Reclaimed water pipe has been incorporated into all new whole barracks renewal projects since FY02, allowing the reuse of rainwater for the facilities' nonpotable water needs. Fort Lewis has also installed rainwater cisterns in a major new barracks project and participated in regional watershed planning, as we will discuss in greater detail below in the community and regional interaction subsection.

---

[26] Information provided by Fort Hood staff.

Products and Materials Management

A main focus of many installations' sustainability programs has been the sustainable procurement and disposal of products and materials. Purchasing programs have focused on more sustainable, "green purchasing" in everything from office supplies to furniture and building construction materials. As mentioned earlier, Fort Lewis pioneered the SIS, which has resulted in the purchase of more than $180,000 in recyclable and/or recycled content furnishings (see Box 2.4).[27] A couple of other installations have started to implement SIS programs, such as Fort Jackson, which had its first SIS demonstration displays during summer 2007.

Fort Carson has used its green purchasing program to invest in more energy-efficient, high-quality washers in soldiers' barracks. The Directorate of Logistics purchased 224 ENERGY STAR[28] qualified washers for the 14 renovated "rolling pin" barracks. These frontloading washers use less water and energy and have lower maintenance costs.[29]

Many installations have accomplishments in recycling, reuse, and process changes to reduce the use and disposal of solid waste and hazardous materials and waste. Strong pollution prevention programs, cost savings, and concerns about hazardous materials and available landfill space have contributed to these successes. Fort Hood, for example, saved more than $2.5 million in 2006 through its qualified recycling program, compost recycling program, inert material management, deconstruction management, special waste management, and electronics waste recycling program.[30] Fort Hood has an extensive program to recycle and reuse solid wastes in its "Every Waste a Reuse Opportunity" program. (See Figure 3.5.) Materials recycled and reused include aluminum, plastics, wood, metals, batteries, paper, cardboard, asphalt, mattresses, soil, tires, motor oil, and antifreeze. In FY06, Fort Hood recycled a total of 49,715 tons of solid waste, generating a total revenue of $1,738,778.[31] Similarly, combined solid waste reduction initiatives saved Fort Carson $500,000 in FY06.[32]

To address hazardous wastes, Fort Hood has changed its vehicle-cleaning processes so they use only one solvent type, and then almost all of that solvent is reused to avoid hazardous wastes and their disposal costs. This process includes a distillation

---

[27] Fort Lewis, 2006.

[28] ENERGY STAR is a joint program of the U.S. Environmental Protection Agency and the U.S. Department of Energy to provide labeling for energy-efficient products. It is a voluntary labeling program designed to identify and promote energy-efficient products to reduce greenhouse gas emissions. The ENERGY STAR label is now seen on over 50 product categories, including major appliances, office equipment, lighting, and home electronics. For more information see: http://www.energystar.gov/

[29] Fort Carson, 2006, p. 10.

[30] U.S. Army, June 13, 2007.

[31] Fort Hood, 2006.

[32] Fort Carson, 2006, p. 11.

Figure 3.5
Fort Hood Solid Waste Recycling Center

SOURCE: Photo by Beth Lachman.
RAND *MG837-3.5*

system to recover remaining solvent from drums, so that the post can reuse the drums and use nearly all the solvent.[33]

Like Fort Hood, other installations have accomplishments related to solvents and equipment and vehicle-cleaning processes. For example, Fort Benning's Centralized Hazardous Material Control Center (CHMCC) provides on-site solvent recycling for parts washers and weapons cleaners located throughout the installation. Recycling saves approximately $196,000 per year in avoided disposal costs.[34] Similarly, Fort Rucker's Aviation Center Logistics Command has created a program with a local industrial laundry to recycle absorbents used to wipe aircraft engines, which eliminates hazardous waste while also reducing aircraft cleaning costs. The absorbents were previously discarded as hazardous waste after one use due to the presence of the toxic metal cad-

---

33 Information provided by Fort Hood staff.

34 Fort Benning, February 15, 2007, p. 102.

mium. The absorbent material can now be reused as many as ten times before being discarded. Fort Rucker's staff estimates this approach to save about $500,000 a year.[35]

Another major area of accomplishment is in reusing materials and waste from construction and deconstruction activities, which save money and landfill space. Forts Bragg, Campbell, Hood, Lewis, and Carson all have programs to grind and reuse concrete and to deconstruct buildings rather than demolishing them. Below we present several examples from Forts Lewis, Campbell, and Carson that illustrate these accomplishments and the savings achieved.

In 2006, Fort Carson diverted large quantities of construction debris and pallet wood totaling about 40 tons per month from a landfill and sent them to a contractor who turned the waste into mulch. More than 80 percent of the construction debris, by weight, from the "rolling pin" barracks renovation project was diverted from a landfill through recycling and reuse efforts.[36] By June 2007, Fort Lewis had diverted more than 725 tons of organic material and 1,400 tons of waste wood from its solid waste stream and avoided $174,000 in disposal costs by reusing lumber and other resources from building deconstruction.[37]

Fort Campbell and Fort Hood have both sold buildings to the highest bidder for deconstruction and material salvage instead of demolishing them. For example, at Fort Campbell one World War II era building was deconstructed for a cost savings of $5,135 relative to the cost of demolition; much of the cost savings came from avoiding landfill costs.[38]

Fort Campbell, like several other installations, successfully grinds concrete from demolished structures into gravel for reuse. (See Figure 3.6.) Concrete aggregate is then used for various purposes around the installation. Demolition on major construction projects requires on-site concrete grinding for possible reuse during site improvements, etc. Grinding concrete results in a significant increase in waste diversion from demolition projects. This process costs $5.86 per ton of concrete and saves $30 per ton, resulting in an annual cost avoidance of $600,000, as of September 2007.[39]

Natural Resources Management

Installations often do not include much natural resource activities, such as species, habitat, and ecosystem management activities and biodiversity protection, in their sustainability programs, even though installations are performing these activities. This frequently occurs because pollution prevention and environmental engineering staff

---

[35] U.S. Army, June 13, 2007.

[36] Fort Carson, 2006.

[37] PR Newswire, June 3, 2007.

[38] Information provided by Fort Campbell staff.

[39] Fort Campbell, undated (a).

Figure 3.6
Fort Campbell's Concrete Grinding Creates Reusable Gravel

SOURCE: Photo courtesy of Fort Campbell.
RAND *MG837-3.6*

run the sustainability programs, and installation natural resource staff are not very deeply involved in the sustainability program. The result is that sustainability programs do not report many accomplishments in this area.

We found a couple of exceptions to this pattern. The best example, as discussed in the last chapter, is Fort Lewis. Fort Lewis's natural resource accomplishments include developing and distributing oak woodland and prairie management plans; locating and abating invasive species, such as eradicating 1,581 acres of scotch broom and other unwanted vegetation; experimenting with planting and growing native species that are host plants for rare butterflies; implementing sustainable forestry within its timber program; and collaborating with Washington state and other federal employees to develop a regional protocol for defining quality of prairie habitat. The Integrated Training Area

Management greenhouse program staff has collected native seeds from Fort Lewis training lands and is propagating 32,400 native plant plugs.[40]

Fort Benning was trying to integrate some natural resource activities into its sustainability program, but its accomplishments in this area within the ISP were limited in 2007 because its ISP program was relatively new.

Fort Carson has incorporated some of its natural resource activities into its sustainability program. Its accomplishments include starting sensitive species studies, such as an antelope fawn survivability study at Pinyon Canyon Maneuver Site (PCMS) and a study on burrowing owls, a Colorado threatened species, that was initiated in 2006 at PCMS and has identified 30 nesting pairs and banding young. Strategically participating in the Central Shortgrass Prairie (CSP) ecoregion partnership to help prevent the black-tailed prairie dog and other declining sensitive species from becoming threatened and endangered is also included in its sustainability activities.[41]

Other installations have some limited natural resource activities incorporated into their ISPs, and thereby limited accomplishments. These include the ACUB conservation buffering activities and habitat restoration projects that directly benefit training ranges, such as erosion control projects. It is important to note that many of these installations have strong natural resource programs and accomplishments in these areas, but they are not integrated into the sustainability program.

Community/Regional Interaction

Some installations are trying to reach outside the fenceline to address broader community and regional sustainability concerns that affect the post, such as encroachment and regional transportation issues. Many have active ACUB programs, such as Forts A.P. Hill, Benning, Bragg, Carson, Lewis, and Stewart, which they may or may not consider part of their sustainability programs. Most of these programs are making progress in protecting land with conservation easements to help prevent encroachment. However, most installations are not yet communicating and collaborating enough with external communities on sustainability, especially regarding more strategic issues such as regional transportation and growth management. Community and regional collaboration takes time and resources, particularly manpower for outreach activities. It is important to note that since this research was done in 2007, many of the later ISPs have

---

[40] Fort Lewis, 2006, and Fort Lewis, 2004.

[41] Fort Carson includes prime habitat of the CSP ecoregion, which encompasses approximately 56 million acres and includes parts of Colorado, Kansas, Nebraska, New Mexico, Oklahoma, Texas, and Wyoming. The CSP contains 146 animal and plant species that are state and/or federally listed and/or are considered imperiled, endemic, or declining. These species include the black-tailed prairie dog, burrowing owl, mountain plover, ferruginous hawk, and swift fox. The CSP partnership is a collaboration of different federal, state, nongovernmental organization (NGO) actors and private landowners to study, manage, and preserve the CSP ecoregion and maintain a healthy, viable ecosystem so that species do not become threatened or endangered. For more information, see Neely et al., November 2006.

been addressing community and regional interaction more in their development and implementation process. For example, Fort Detrick, the California National Guard, and Letterkenny Army Depot have been looking at off-post transportation issues and at community/regional partnerships to address them.

Some installations are attempting to include local communities in sustainability planning, but do not have enough staff and other resources to conduct sufficient outreach and collaboration with the communities. This type of collaboration is difficult in more rural areas that are more interested in economic growth than in strategic, long-term sustainability issues. For example, Fort Bragg has given funding to Sustainable Sandhills, a regional sustainability planning collaboration,[42] and is trying to promote regional growth management. One of Fort Bragg's sustainability objectives is to "Initiate and lead state and regional planning forums."[43] Fort Bragg is also working with the DoD Office of Economic Adjustment and surrounding communities on regional growth issues related to 2005 BRAC directives. However, not much has yet been accomplished by any of these activities. In addition, many local counties do not yet understand sustainability and the importance of collaborating on growth management and other regional sustainability issues.

Fort Jackson had also tried to focus on regional interaction and collaboration. One of its five core teams is focused on regional interaction, and it has a goal focused on "Integrated compatible regional land use." This installation formed the Midlands Area Joint Installation Consortium and a low-impact development working group to define projects to look at larger ecosystem impacts and requirements. However, since these efforts were fairly recent in 2007, not much had been implemented yet beyond this collaborative planning and analysis process.

Fort Carson has demonstrated some initial accomplishments in community outreach in a few areas. Fort Carson hosts a large sustainability conference every fall with the community and is trying to do some transportation planning coordination. In partnership with the Pikes Peak Area Council of Governments (PPACG), Fort Carson has initiated the Pikes Peak Sustainability Indicators Project, which is being integrated with the BRAC implementation planning process. Measures or indicators will be used to objectively determine whether the region is performing in a sustainable way. The PPACG, local governments, and Fort Carson will then partner, collaborate, and coordinate strategies and initiatives if an improving trend is desired. In a related activity, Fort Carson is developing a "Fort Carson Regional Growth Coordination Plan." However, so far there is no regional growth planning among neighboring communities, and many local governments promote suburban sprawl under their current policies. Local governments provide some support to the installation to reduce nearby sprawl because

42 For more information, see the Sustainable Sandhills web site, "Our Mission and Our Vision," at http://www.sustainablesandhills.org/.

43 See Fort Bragg, May 8, 2006.

of encroachment concerns, but are not yet supporting broader sustainability issues such as regional growth management.

Fort Lewis has a few accomplishments in community collaboration, mostly related to natural resources. It has been working with local governments on watershed planning. For example, in September 2006, Fort Lewis conducted a town hall meeting for Fort Lewis and surrounding communities to discuss the Murray Creek and Sequalitchew Creek watershed management plan. It has also been working with the Washington state departments of Natural Resources (WDNR) and Fish and Wildlife (WDFW) on species and habitat management issues. For example, Fort Lewis and the WDFW developed an agreement to coordinate and implement recovery efforts and plans for South Puget Sound rare species.[44] Another example is that Fort Lewis is using some ACUB funding for research and prescribed burns to help restore prairie habitat on WDNR land at Mima Mounds Natural Area Preserve (see Figure 3.7).[45]

Quality of Life Issues

Most installations do not include many quality of life (QOL) issues, such as health, education, and family support concerns, in their sustainability programs. Many staff that we interviewed did not even understand what QOL meant for an installation and which installation functional areas affected QOL issues. A review of the sustainability literature and sustainable community activities suggests some key QOL areas:

- Economy, wealth, and employment
- Housing and community facilities
- Land use and density
- Education and youth involvement
- Health and wellness
- Communications (e.g., internet access)
- Recreation and culture
- Crime and safety
- Civic involvement and governance
- Spiritual opportunity/participation

Not all of these are directly relevant to the Army ISP process, but they provide a useful starting point for installations to learn and think more about QOL issues. Examples of QOL concerns for Army installations include whether soldiers and their families have quality housing, transportation, healthcare, education, and recreational opportunities. Do they feel safe, happy, and satisfied with their jobs, incomes, homes,

---

[44] Fort Lewis, 2004.

[45] For more information about the Mima Mounds Natural Area Preserve and prairie habitat in the South Puget Sound, see the web site at http://www.southsoundprairies.org/visit.htm.

Figure 3.7
Mima Mounds Natural Area Preserve, Washington State Department of Natural Resources

SOURCE: Photo by Beth Lachman.
RAND *MG837-3.7*

commutes, community, and lives? Such concerns can affect soldier retention and job performance and are important aspects of sustainability. Similar QOL concerns can also apply to other installation employees, military retirees, and the broader community. See Appendix B for more information about the role of QOL in sustainability. Since this research was done in 2007, more ISPs are addressing such QOL issues. For example, ISPs initiated late in FY07 and early FY08 (such as USAG Wiesbaden, Letterkenny Army Depot, and Fort Detrick) have been looking at education, civic involvement, Army Community Services (ACS) and MWR services, such as providing recreation and chapel facilities.

Given that most installation sustainability programs receive funding and support from the installation's environmental program, it is understandable that QOL issues were not being addressed by most sustainability programs. This disconnect has also arisen because the ISP development process tends to focus more on environmental technology and other traditional business and industry environmental areas. The sustainability consultants who help facilitate the Army's ISP development process have mostly been engineering and environmental technology experts with little experience in sustainable community activities that focus more on QOL issues. Therefore, there are few sustainability accomplishments in this area.

Three installations that had incorporated some QOL concerns into their sustainability programs are Fort Benning, Fort Jackson, and Fort Carson. As discussed in the last chapter, Fort Benning includes QOL issues, such as improving family and soldier support and designing facilities to address well-being concerns, in its goals, objectives, and actions. However, since Fort Benning's sustainability program is relatively new, there were limited accomplishments as of fall 2007. Plans have been developed to address QOL concerns through upgrades in service support facilities and activities, such as the Commissary, recreation centers, building a new hospital, and addressing education needs, but none of these projects has yet been completed.

Fort Jackson was developing a "dashboard" of QOL indicators as part of its sustainability program. PAIO staff were trying to integrate them into the installation strategic plan. However, since Fort Jackson's sustainability program was also relatively new, there were limited QOL implementation accomplishments associated with this effort as of fall 2007.

Fort Carson had a couple of sustainability activities that address QOL issues and has made some progress. The Pikes Peak Sustainability Indicators Project mentioned above involves counties, municipalities, and other stakeholders in the tracking of long-term trends for select QOL measures. A Citizens Soldier Connection Program was created in 2006 to connect Fort Carson soldiers and their families with volunteer citizens in the community to help provide a local support network. The objectives of the program are (1) to connect soldiers with Colorado Springs/Pueblo community members; (2) to create positive off-duty interactions for all participants; and (3) to develop potentially enduring relationships. By fall 2006, 300 matches had been made.[46]

As discussed earlier, there also were some limited QOL accomplishments related to the design of facilities, buildings, and housing areas, which are focused more on soldier and family QOL concerns. However, much of the progress in building design was due to the Army's green building requirements. Exceptions are Fort Bragg's barracks campus designs and Fort Carson's Town Center concept, which are more strategic master planning activities and integrate more QOL issues. These efforts show some initial progress in addressing QOL concerns related to community living spaces.

It is important to note that installations may have QOL accomplishments, such as family support programs, but these activities are not being integrated into the installation sustainability program.

---

[46] Fort Carson, 2006, p. 4.

Clearly, installations have made progress in implementing sustainability, but some gaps remain. In the assessment of the implementation process in the next chapter, these gaps are discussed in greater detail.

CHAPTER FOUR

Assessing the Needs of the ISP Development and Implementation Process

To produce HQDA guidance that will help more installations develop and implement effective ISPs, it was important to analyze both the development and implementation processes. This section describes those assessments.

Assessment of the ISP Development Process

As was described in Chapter Two, up to and partway through FY07, ISPs were being developed based on a process model consisting of a series of multiple workshops. Here, we discuss the advantages and disadvantages of this process, as well as what is needed to improve it. It is important to note that this process changed in 2007 and is no longer the current ACSIM process. In 2009 the process includes fewer workshops and focuses more on the local needs of the installation, as recommended here.

Advantages of the Early 2007 ISP Development Process

The ISP development workshop series helps to create a foundation of installation staff interest in sustainability. First, the workshops educate a wide range of people at an installation about sustainability, some of whom have never been exposed to the concept before. Second, they motivate diverse installation staff to participate and collaborate in the sustainability process. Staff from different functional areas, such as transportation, logistics, contracting, and environmental management, have a chance to meet, develop relationships, and begin to collaborate on sustainability. Third, this process helps to create sustainability champions, individuals who are inspired and motivated to implement sustainability.

The workshops help staff to develop an ISP and start implementing sustainability projects. The workshop series helps the installation develop some initial sustainability goals, objectives, and teams for the ISP. The workshops also help demonstrate that the garrison commander values sustainability, so installation staff see that sustainability is important.

Disadvantages of the Early 2007 ISP Development Process

However, there were also some disadvantages to the multi-workshop-based ISP development process. First, it was time consuming. It could take a year or more to hold all the workshops. Second, it was expensive. It cost approximately $130,000 to $150,000 to conduct the series of workshops, not including the time spent by installation staff attending the workshops. Some costs, such as travel costs, can be reduced by combining workshops, as was done at USAG Hawaii. However, this step did not reduce total costs very much. Costs are high because the workshops are manpower intensive. Two highly trained facilitators are needed for each sustainability team, so the process typically required 10 to 12 facilitators at each workshop, including both Army and contractor personnel. Given the number of trained facilitators available, the four-workshop process was impractical if the Army wanted a significant number of installations to develop ISPs over the next couple of years. It was too expensive, time consuming, and manpower intensive.

Third, the initial ISP goals and objectives developed by the workshop process often missed key areas because they depend on the workshop focus, outcomes, and the participants who attended. Overall, the process tended to focus more on "business environmental" activities and easier objectives with obvious economic benefits because of the structure and the sustainability examples presented at the workshops. Many of the sustainability presentations at the plenary sessions also tended to address broad environmental and sustainability concepts rather than the specific mission needs and circumstances of the installation. QOL, mission areas (other than training and related environmental issues), regional collaboration, and ecosystem concerns were not as fully addressed by the process. Key participants, such as health, MWR, and other QOL staff, may not be invited or may not attend the workshops. Such staff are needed to bring more QOL ideas into the process.

Fourth, there was no analysis to identify the most important, strategic, or pressing sustainability issues. For example, it may be important to focus on addressing encroachment through conservation buffering before nearby development occurs, but that may not be a priority because no analysis was done or participants were not aware of this need. Workshop participants were often thinking off the top of their heads, and may only be familiar with issues related to their functional areas. A voting process, rather than objective analysis, was used at the workshops to prioritize sustainability objectives. The former is likely to have effectiveness problems if all participants do not have broad knowledge of the sustainability issues.

Fifth, there was no guarantee that the workshop process will actually lead to the development of an ISP document and the implementation of the ISP. A few installations that went through the workshop series have either not written an ISP or have not done much to implement their ISP. The lack of follow-through can occur for a number of reasons, such as a lack of resources, time, or commander and technical support after the workshop series ends. In some cases, it may be due to the lack of a requirement to produce an ISP document. An installation may even be implementing a sustainability program without an ISP document. For example, Fort Jackson had been implementing sustainability projects and other activities since its workshop series ended, but had not written an ISP document because its sustainability staff were busy with implementation activities. It is important to note that while these earlier workshops did not always include the development of an ISP, all of the later planning culminated in "action planning" sessions that produced a standalone ISP or sustainability goals, objectives, and actions (and metrics to track progress) incorporated in the installation's strategic plan.

ISP Development Needs

Given the disadvantages of the FY07 workshop process, the ISP development process needed to be streamlined so that more installations can develop ISPs in a more timely fashion. Workshop presentations on sustainability should be customized more for the mission, QOL, and environmental needs and concerns of each individual installation. There is a need for more technical support to ensure that installations develop an ISP document, in addition to workshop support. Policies should be instituted to require installations to develop an ISP document after the workshops end, for all key functional staff to be involved in ISP development, and for specific functional areas to be considered as part of the sustainability planning process, including key QOL areas such as health, education, and MWR. As noted earlier, since late 2007, ACSIM has made significant changes to streamline the ISP process and address many of the issues identified here.

Assessment of the ISP Implementation Process

In the last chapter we discussed the progress that installations have made in implementing sustainability. We found that many installations are making progress, but some issues and challenges remain. In this section we summarize the issues and challenges and suggest some paths for addressing them to enable more progress in ISP implementation. Based on the last chapter's discussion of accomplishments, we begin by assessing the overall accomplishments in ISP implementation and identifying gaps between accomplishments and ISP intent. Many of the points that we make in this assessment are supported by the business and environmental management and the sustainability literatures.[1]

---

[1] Useful references include Champy, 1995, Kotter, 1996, Kotter et al., 2006, Global Environmental Management Initiative, October 25, 2006, and Lachman et al., 2001. Another useful reference is GEMI's home page at http://www.gemi.org. GEMI, the Global Environmental Management Initiative, is an organization of innovative companies dedicated to fostering global environmental, health, and safety excellence through the sharing of tools and information. This organization has also developed tools for addressing sustainable development. Later in this

Assessment of ISP Implementation Accomplishments and Gaps

Installations clearly have demonstrated some accomplishments given limited guidance, staff, funding, and time. However, some important issues remain. Most installations' focus and successes have been in more traditional business strategic environmental activities, such as implementing pollution prevention and more energy and environmentally preferred technologies; within the mission area, most programs have focused on training ranges. Given the funding sources of most ISP activities, this is not surprising. Most start by focusing on the "low-hanging fruit" in these areas, because it is often easier to implement an environmental technology project that creates some life-cycle cost savings by reducing energy, water, or solid/hazardous waste disposal costs. Often such activities are good starting places to make some initial progress, to get buy-in, to educate diverse installation staff about sustainability, and to help expand the program. These projects also leverage off strong installation pollution prevention, energy management, environmental management systems (EMSs) (such as ISO 14001 implementation), and environmental management programs.

It is likely that at some installations, some of these accomplishments, or even many of them, would have been achieved without a sustainability program. It is difficult to assess and quantify the value added from the formal ISP. However, the value added can be seen more clearly in the staff attitudes and their activities that are trying to address the integration of mission, environment, and community. Installation staff are trying to do more systems thinking and integrating sustainability principles across different functional areas. Fort Bragg's master planning and building construction examples that address QOL and environmental concerns illustrate this point. This systems approach, which considers the interrelationships, integration, and synergies across mission, environment, and community, is a key strength and benefit of a sustainability program. However, there have not yet been enough approaches and activities based on systems thinking that fully address such interrelationships in most installation sustainability programs.

There also are some key sustainability areas that have not received enough attention yet and therefore have few accomplishments. Since there has been less focus on education, health, family support, and QOL issues, there are fewer accomplishments in this area. It is important to note that installations may have successful programs in these areas that are run by other functional organizations, but such activities are not considered part of the sustainability program. Similarly, there is less attention and fewer accomplishments related to conservation, ecosystem management, biodiversity protection, and other natural resource issues. Again, installations may have accomplishments in these areas, but they are not considered part of the sustainability program. Regional collaboration and community outreach on sustainability, especially related to key stra-

chapter we provide some additional references to such literature in footnotes for readers who want more detailed information on specific topics.

tegic issues of regional transportation, growth management, regional water, habitat, and air quality, have not received much attention in most installation sustainability programs. Nonrange mission areas, such as mobilization and deployment, also have received less attention, and thus there are fewer accomplishments in these areas.

Some installations' sustainability processes have evolved over time as program participants have revisited goals and even restructured teams in some cases. These changes are important, since sustainability is considered an evolutionary process. Often such changes were made because installation sustainability staff realized they had left key functional staff and areas out of the process, as occurred at Fort Bragg. Some installations are also starting to evolve their programs to address more complex and strategic issues, such as regional growth.

However, so far most installations tend to place too much emphasis on easier sustainability approaches, such as technology adoption and issues internal to the installation, and not enough emphasis on assessing and considering regional trends, relationships, and effects that can be important for long-term installation sustainability. For example, Fort Bragg has made significant efforts to address water-quality concerns over the years, but managing regional trends may be more important to address future concerns about water quality. In fact, the USACE Construction Engineering Research Laboratory's Strategic Sustainability Assessment (SSA), a pilot study of installations in the fall line region of the southeast, found the following:

> Regardless of urbanization practices on-post, Fort Bragg is likely to experience significant degradations to its water quality due exclusively to practices within the surrounding communities. Sustaining water quality on-post and within the local communities requires that decision makers, planners, developers, special interests, and politicians perceive their communities as part of a larger system, with the success of any single component dependent on the success of the system.[2]

The Construction Engineering and Research Lab (CERL) analysis indicates that water quality issues at Fort Bragg need to be addressed by a broader and more strategic regional planning and systems approach in collaboration with local communities.

In addition, installation sustainability project activities are not being analyzed or documented well. Most of the project examples we cited, even from written reports, had no thorough analysis of the true costs and benefits to the mission, environment, and community. Full life-cycle cost analyses that address all labor and material costs and mission effects, and consider full cradle-to-grave implications of the activity for the environment and community both on and off post, are lacking. For example, an assessment of Fort Bragg's urban village construction should include the range staff man-hour investment in the project relative to the traditional man-hour investment; a full materials investment comparison; an assessment of the training benefits from having

---

2 Jenicek, Myers, et al., November 2006, pp. 202–203.

the urban village five years earlier; and an analysis of the effect on total waste sent to the landfill. Another example is that the full worldwide environmental and sustainability benefits and costs of using E-85 fuel should be assessed. Such assessments are needed to truly measure and demonstrate the impacts of these projects and to provide information about which projects are most effective.

There also is no consistent, up-to-date documentation about installations' sustainability activities. Installations do not have a requirement to document what they are doing and what has been accomplished, and most do not spend the time to do it. Many installations lack clear metrics linked to objectives and goals and ongoing tracking of those metrics. Information that exists is mostly in PowerPoint charts from status presentations and updates to senior management about the sustainability program by goal team experts without consistent metrics being tracked from year to year. Some installations maintain a tracking system in a database (Excel, ACCESS, web-based, etc.) to document success at meeting sustainability objectives, tasks, and initiatives. However, without a true assessment to measure progress and without written documentation about project accomplishments, it is difficult for HQDA and other staff outside the installation to see, measure, communicate, and spread the benefit from these accomplishments. It is important to note that since this research was done in 2007, later ISP planning activities are including more metrics and improving their reporting.

Installations have yet to integrate sustainability into core business processes, which is key to implementation. Some have integrated sustainability into their environmental management systems, which is a good step to integrate sustainability into key environmental processes. Some have integrated sustainability into installation strategic plans and other core planning documents, but more needs to be done in other information management systems and installation processes. In addition, none appear to be integrating sustainability into Lean Six Sigma programs. Bringing business process reengineering principles into sustainability planning could generate benefits to mission, community, and the environment as well as life-cycle cost savings, given the potential synergies and complements between the two programs.

Challenges to ISP Implementation

Installations have accomplished quite a bit considering the challenges they face in trying to implement ISPs. First, many installations do not have sufficient resources, including funding and staff, to implement broader sustainability programs. Sources of funding for sustainability projects are currently limited. Most funds come from installation environmental programs, with some additional funds from sustainable range programs. Since these funds are primarily for environmental programs, it is difficult for installations to implement non-environment-related sustainability projects, such as QOL projects.

Allocating sufficient staff time to sustainability is important for implementation. Installations with a dedicated sustainability coordinator are able to accomplish more than those that do not. However, that is not enough; having other functional staff who can devote time to sustainability is key as well. Involving other functional staffs, such as strategic and master planning, PAIO, training range, MWR, and health staffs, helps to integrate sustainability into different installation functions and to address a broader range of sustainability concerns. Many staff members who are assigned to sustainability teams as an additional duty may not have the time to devote to the tasks.

Second, a lack of mechanisms to benefit from return on investment (ROI) also limits ISP implementation. Often, sustainability projects require an upfront investment to achieve future benefits, such as lower energy costs or less water usage. Efforts may be further complicated if additional upfront costs occur in one budget account, such as higher military construction costs to meet LEED Gold or Platinum standards, but savings occur in other accounts, such as reduced operations and maintenance costs for energy, water or wastewater disposal. For example, Fort Carson's 2006 sustainability report mentions that being able to find and invest "capital to support higher first cost of sustainable construction in order to achieve life-cycle savings" is an impediment to ISP implementation.[3] In some cases, the organization building a facility does not want to invest in sustainability features, such as more energy-efficient technologies, that reduce operating costs later. At a couple of installations, USACE construction staff were so focused on reducing construction costs that it hindered LEED implementation, even though LEED is an Army standard. At one installation, USACE staff claimed that LEED was not applicable, and at another, LEED features were cut out to save money as construction proceeded, because the project costs started to overrun the budget.

In other cases, a sustainability project may cost more than the standard approach, but there are significant environmental, operational, or QOL benefits that are not easily quantified, making the project difficult to justify. For example, incorporating a LID project when building a parking lot may improve water quality and address stormwater runoff problems, but the organization responsible for building the parking lot may not want to pay the extra cost. Another ROI problem is that newer, more sustainable technologies may require higher maintenance costs to achieve the intended benefits. If maintenance funding is not increased, some of the benefits of the project may be lost.

Annual budgets are often based on prior-year spending or standardized cost models, such as OSD's Facility Sustainment Model and Facility Modernization Model or the Army's Standard Service Costing and Service Based Costing.[4] Therefore, it may be difficult for installations to obtain the necessary funding for higher upfront costs of sustainable products and services, unless they cut costs or services elsewhere. Other problems noted by installation personnel include difficulty keeping revenue from recy-

---

[3] Fort Carson, 2006, p. 8.

[4] See, for example, Unisys Corporation, April 28, 2005.

cling or retaining savings to reinvest in further sustainability efforts, except in areas such as Non-Appropriated Fund accounts or the Residential Communities Initiative.[5]

Third, there is insufficient experience, education, and training for staff implementing sustainability projects. Sustainability knowledge is required by diverse installation staff, not just the sustainability coordinator. The sustainability coordinator, range staff, DPW, PAIO, and other key staff need education about sustainability implementation. A couple of installations, such as Fort Carson and Fort Hood, have developed and implemented sustainability training for installation staff. However, most installations do not have the staff, time, or expertise to provide such education and training. The problem is that after the ISP development workshops end, most installation staff are on their own; there is no further Army-wide sustainability training. Given the lack of consistent, up-to-date documentation, there is also limited transfer of information from lessons learned and experts at other installations. Therefore, it is difficult to transfer knowledge and success stories across installations. Only informal lessons and anecdotes are being shared.

In some cases, there may be misinformation about Army or DoD policies related to sustainability. For example, some installation staff cited the Federal Acquisition Regulation (FAR) as an impediment to greener procurement. Though there may be some problems in specific instances, the FAR has some subparts supporting more sustainable procurement. Examples include requirements that federal agencies purchase many types of products made with recycled materials through Affirmative Procurement Programs; acquire energy- and water-efficient products and services, and products that use renewable energy technology; and implement cost-effective contracting preference programs promoting the acquisition of environmentally preferable products and services, eliminating or reducing the generation of hazardous waste, and minimizing the procurement of materials and substances that contribute to the depletion of stratospheric ozone.[6]

Fourth, many installations are having the problem that sustainability is viewed as an environmental program. Some installations, such as Fort Hood, are working to integrate sustainability into all installation functions, but it is difficult to do this when some installation staff view sustainability as an environmental or DPW role, and thus

---

[5] See, for example, Fort Campbell, September 2003. The Residential Communities Initiative involves public-private partnerships to renovate and replace Army family housing on installations. Private-sector contractors are better able to trade off higher upfront construction costs for future savings.

[6] Federal Acquisition Regulations Subpart 23.4—Use of Recovered Materials, Subpart 23.2—Energy and Water Efficiency and Renewable Energy, Subpart 23.7—Contracting for Environmentally Preferable Products and Services, and Subpart 23.8—Ozone-Depleting Substances can be seen at the FAR web site. As of October 2, 2007:
http://acquisition.gov/comp/far/current/html/Subpart%2023_4.html
http://acquisition.gov/comp/far/current/html/Subpart%2023_2.html
http://acquisition.gov/comp/far/current/html/Subpart%2023_7.html
http://acquisition.gov/comp/far/current/html/Subpart%2023_8.html

something they do not have to worry about, since the environment division runs the sustainability program. This attitude makes it more difficult to integrate sustainability more broadly and to address nonenvironmental sustainability issues, especially QOL issues. When the sustainability program is located in the post's environmental organization, it contributes to this perception. Often, nonenvironmental staff do not see a need to participate, since they think the environmental staff will do it.

In other cases, installations have difficultly implementing sustainability, because it is viewed as just another management "flavor of the month" and a passing fad that will disappear in a year or two when management changes occur, as Total Quality Management (TQM) and Lean Six Sigma are sometimes viewed. Installation staff who resist change and are used to doing things the same old way resist implementation of the ISP. At one installation, the installation sustainability coordinator said, "I can't tell you how many times I have heard someone say, 'I can outwait you.'" This sustainability coordinator was quite frustrated because such staff ignored and even tried to stop sustainability activities.

Fifth, many installations do not always have sufficient senior management support. Garrison commanders change jobs every 2 to 3 years, which can slow the momentum to implement an ISP. A new commander often does not know about sustainability or appreciate the wide range of benefits an effective ISP program can have. Installation directorate leaders, such as the heads of DPTM and DRM, also need to support sustainability. Sustainability staff from three different installations stated that they had a garrison commander who provided "lip service" support, such as at public relations events, but did not require certain key directorates to participate in the program. Several installations' sustainability staff were frustrated because senior staff did not provide adequate support in terms of resources, multifunctional team participation, and other support needed to effectively implement an ISP. A lack of senior management support was part of the reason many programs were not able to address sustainability issues in areas outside DPW and environmental management. For example, at a couple of installations, senior management did not support efforts to integrate sustainability into the master plan and installation strategic plan.

Sixth, implementation of ISPs can be limited by installation and Army policies, practices, and guidance that do not yet include or may impede sustainability. For example, USACE design standards, such as those for barracks, ranges, and chapels, can slow down or limit sustainability implementation. When Fort Bragg wanted to build barracks six stories high, it had to get an exemption from the USACE barracks standard, which required barracks to be fewer stories. Another installation found that BRAC implementation guidance was making it difficult for them to implement sustainability in new facilities. Another example is Army Regulation (AR) 25-50, "Preparing and Managing Correspondence," which requires installations to use one side of the paper. This regulation made it difficult for one installation to implement a sustainability policy to use both sides of the paper to reduce paper use.

When installation staff can point to official Army guidance and policy that supports sustainability, it is easier to implement. For example, since LEED is an Army standard for new construction, it is easier for installations to incorporate more sustainability elements into new building construction. However, as indicated by the LEED implementation problems we discussed above, having a policy to support sustainability does not necessarily mean it is always implemented properly, especially when more sustainable approaches incur additional up-front costs.

Factors That Facilitate ISP Implementation

Based on our discussions with installations about their sustainability programs, our review of the management and sustainability literature, and making comparisons across installations, we identified eight factors that facilitate ISP implementation.

First, having at least a half-time sustainability coordinator is important to help manage and implement the ISP, to provide education and technical support, and to motivate others to participate in the sustainability program.

Second, having enthusiastic sustainability champions helps with ISP implementation. Since implementing sustainability is very much about instituting change, knowledgeable sustainability "champions" are needed to help educate and motivate others to implement sustainability.[7]

Third, support from the garrison commander and other senior management is needed for more efficient and effective sustainability implementation. Installations where the garrison commander and his directorates viewed sustainability as important had wider support, more team participation, and fewer difficulties in implementing sustainability. Installations are more successful at implementing sustainability when the garrison commander clearly supports the effort.[8]

Fourth, having sufficient, consistent, and diverse funding sources for sustainability activities helps with ISP implementation. Installations with ongoing support for a sustainability coordinator and a range of projects were able to accomplish more. Looking across all the ISP activities over time, installations have acquired funding from DPW environmental programs, sustainable ranges programs, the garrison commander, FORSCOM, IMCOM, and ACSIM. Having access to funding from these multiple sources helps with sustainability implementation.

Fifth, integrating sustainability principles and practices into other installation planning and operational documents helps promote ISP implementation. Some installations have successfully integrated sustainability into other key installation documents,

---

[7] For more information from the business and environmental management literature about the importance of champions and building a coalition of support for change, see Kotter, 1996, and Lachman et al., 2001, p. 66.

[8] The business and environmental management and the sustainability literatures stress the need to have management support for integrating approaches like sustainability. For example, GEMI's Sustainable Development Planner stresses the need to have senior and key middle management commit to sustainable development, see GEMI, October 25, 2006, p. 14.

which helps integrate sustainability into all installation functions.[9] Fort Bragg's integration of sustainability principles into its master plan and IDG helps it design more sustainable building and facilities. Fort Hood's training lands team has developed a Land Sustainment Management Plan (LSMP) as part of its sustainability program and has integrated parts of this plan into its INRMP. Such integration also helps these installations implement sustainability activities because they can point to the official guidance and policy when they encounter people who are unaware or uninterested in sustainability implementation.

Sixth, including sustainability as a performance evaluation criterion for managers and staff is another important factor for ISP implementation.[10] When employees are measured and judged based on their performance in an area, they pay more attention to that area. At most installations, only the sustainability and environmental staff have sustainability as an evaluation criterion in their performance reviews. One exception was at Fort Bragg, where sustainability is a criterion for some DPTM staff.

Seventh, when installation staff view sustainability as an installation priority, not just an environmental concern, it is more likely to be implemented and integrated throughout installation activities. Fort Bragg is an example where sustainability is viewed by many installation staff, such as range and master planning staff, as an installation-wide program.

Eighth, access to Army sustainability experts and those who have experience implementing sustainability projects helps an installation implement its ISP. Forts Bragg, Lewis, Hood, and Carson all benefited from having FORSCOM sustainability support, such as monthly conference calls, to help them implement their sustainability programs when they first got started. The sustainability coordinator from Fort Jackson visited Fort Lewis and learned a lot from their experts. Similarly, the SIS expert from Fort Lewis has visited several other installations, including a Navy base, to provide expertise about SIS implementation.

Implementation Needs

There are a number of needs that should be addressed if the Army wants more installations to effectively implement ISPs. For discussion purposes, we have grouped them into nine categories.

First, the Army needs to address resource problems, including providing staff for sustainability projects. Sufficient staff time is needed at the installation level to sup-

[9] The environmental management and sustainability literatures stress the importance of integrating the environmental/sustainability concepts into core business processes, and integration into planning and operations is key. In fact, GEMI's fourth step in its Sustainable Development Planner is to make sustainable development part of everyday business processes (GEMI, October 25, 2006, p. 13).

[10] Companies such as Kodak and P&G have used performance evaluation and other incentives to improve environmental management and performance. See, for example, U.S.–Asia Environmental Partnership, October 1997, and Lachman et al., 2001, chapter 7.

port sustainability. Installations need some additional funding for a full-time (or at least half-time) sustainability coordinator, for at least a few years to get the sustainability program started. Many installations probably need a full-time coordinator; as one coordinator stated, "Coordinating sustainability efforts in light of the 'road blocks' is a more than full-time requirement." Without such support, many installations will likely go through the process to develop an ISP but not to implement it. If sustainability is assigned to existing staff as an additional duty, it becomes another burden or "flavor of the month" and staff are not likely to have time to do much even if they want to. An installation needs at least one sustainability "champion" who is dedicated to sustainability and serves to educate and motivate others at the installation. Ideally, once sustainability thinking and approaches have been integrated into installation operations, a sustainability coordinator may no longer be needed. However, until that long-term goal is achieved, installations need to have a sustainability coordinator.

Second, the garrison and unit commanders and senior installation staff, including directorate heads, need to be educated about sustainability and support it. If the garrison commander does not fully embrace sustainability concepts, there is a danger that the program will get lip service only, which hurts ISP implementation. The bosses need to say that sustainability is important. Garrison commanders must instill in each directorate the message that sustainability is important and that staff need to implement it. Unit commanders must instill in troops the importance of implementing sustainability. The Army also needs to include sustainability in garrison and other installation senior leadership performance reviews. If people are evaluated on achieving sustainability results, it motivates them to participate more in the process.

Third, the Army needs to address funding issues and mechanisms for implementing positive ROI programs even when the benefits and costs are split across organizations. HQDA needs to ensure that sustainability investments are being made even if they require higher upfront investments to achieve future cost savings or operational benefits, such as improved environmental performance. Mechanisms need to be developed so that those who invest in more sustainable technologies and systems gain some financial or other benefits from doing so. The Army also needs to make sure that funding is available to provide appropriate maintenance for long-term investments in more sustainable building technologies and other types of sustainability infrastructure and equipment. Since this research was conducted in 2007, ACSIM has been addressing some of these funding issues by including sustainability funding in the Army environmental requirements cost model.

Sufficient sustainability project funding is needed in environmental, mission, and QOL areas. Since Army installations have activities in these areas, some existing funds can be tapped into if the garrison commander and other senior staff support sustainability. In addition, in some cases, by coordinating and communicating with other organizations on post, such as natural resource staff, other relevant activities and sources of funding, such as ecosystem management, can be considered part of and integrated into the sustainability program.

Fourth, headquarters guidance is needed to ensure that installations develop ISPs that include and address all relevant functional areas. Functional areas should include training and other mission functions; facilities and buildings development, management, and maintenance; energy, water, and transportation infrastructure; products and materials management; biodiversity, species, ecosystem, and other natural resource management; community and regional interaction, especially related to land use, resource use, and economic concerns; and QOL. This guidance needs to have sufficient information to identify and encourage installations to place additional emphasis on areas where there are currently implementation gaps, such as QOL and regional collaboration issues.

In fact, numerous people involved in the ISP development and implementation process stated a need for more information about QOL issues. Since the Army ISP development and implementation process has relied so much on environmental engineering and business sustainability experts, practices, and literature, there has not been enough integration of lessons learned from sustainable community efforts, which focus more on QOL factors. The Army needs to incorporate more QOL issues into its sustainability activities. It is important to acknowledge that many installations have other organizations that have implemented some effective QOL activities. However, these QOL activities and organizations should be integrated into the sustainability process to ensure that the ISP is using system thinking and approaches to address and integrate community, mission, and environmental issues over the long term.

To better understand common QOL concerns, Table 4.1 presents some sample QOL sustainability indicators from several different communities that have some potential relevance for installation sustainability activities. For more information see Appendix B, which provides a summary of some of the sustainable community literature on QOL indicators and some additional examples of QOL indicators.

Not all of the QOL indicators in Table 4.1 are directly relevant for installation sustainability planning. However, most are useful for thinking about the installation community, and all are relevant for thinking more broadly about the sustainability of a region, which should be part of the ISP process.

Fifth, the Army needs to better integrate sustainability into installation operations. This is difficult to do without financial support for a sustainability coordinator. It is also more difficult to do in nonenvironmental areas, especially if funding and support for the program comes primarily from environmental organizations. To help facilitate such integration, the Army needs to integrate sustainability into other installation and Army-wide guidance. The Army has started to do this integration with the December 2007 update to AR 200-1, which directs installation commanders to incorporate sustainability principles into the installation strategic plan and other installation management plans.

Table 4.1
Examples of Quality of Life Sustainability Indicators

Category	Indicator	Community
Economy, wealth, and employment	Child poverty and overall poverty rate	Austin[a] Seattle[b]
	Rate of change of median income	Austin
	Share of income by income group (income distribution)	Seattle
	Percent of new jobs filled by local residents	Pikes Peak[c]
	Percentage of total jobs in private sector	Austin
	Unemployment rate	Pikes Peak
Housing and community facilities	Percentage of households able to purchase median-priced home	Austin
	Rental affordability: Percent of households for which average rent would be less than 35% of household income	Austin
	Housing within proximity to commercial and transit services	Pikes Peak
	Family self-sufficiency index (including access to health insurance, cost-effective mobility, daycare)	Pikes Peak
	Streets with bike lanes	Seattle
	Number of community garden plots available	Seattle
Land use and density	People per acre in residential zones (i.e., bedrooms per acre of residential zoning)	Pikes Peak
	Average commute times	Pikes Peak
	Average work commute/vehicle miles	Austin Jacksonville Seattle
	Miles traveled per capita and annual fuel consumption	Seattle
	Annual population change rate	Seattle
Education and youth involvement	Number of students per classroom and per teacher	Pikes Peak
	Percent of all public education students meeting proficiency standards	Pikes Peak
	Annual dropout and graduation rates	Pikes Peak
	Citizen volunteer hours per student in public schools	Seattle
	Satisfaction of parents with the educational system	Pikes Peak
	Adult literacy	Seattle
	Students reporting local or national volunteer service	Seattle
Health and wellness	Health care expenditures per capita	Seattle
	Emergency room use for nonemergencies	Seattle
	Physical activity, smoking, obesity, drug use, sexually transmitted disease, immunizations, and reduction of death due to preventable conditions[d]	Pikes Peak
	Percentage reporting good health status/health care	Austin Jacksonville
	Youth alcohol use	Jacksonville
	Cigarette sales	Jacksonville
	Number of days air quality was good, moderate, unhealthful, very unhealthful or hazardous, according to Puget Sound Air Control Agency	Seattle
Communications	Households with internet access	Pikes Peak

Table 4.1 (continued)

Category	Indicator	Community
Recreation and culture	Tax funding supporting cultural and leisure activities	Pikes Peak
	Percent attending artistic or cultural activities	Austin Jacksonville
	Library circulation and community center visits per capita	Seattle
	Open space/park acreage	Austin Jacksonville
	Areas within 1/8th mile of open spaces	Seattle
Crime and safety	Crime rate	Austin
	Juvenile crime	Seattle
	Reports of family violence or child abuse	Austin Jacksonville
	Vehicle accidents	Austin Jacksonville
	Resident perception of safety and crime statistics, including juvenile arrests	Pikes Peak
Civic involvement and governance	Volunteerism, charitable donations, and individual satisfaction with community involvement	Pikes Peak
	Percentage of registered voters voting in local elections	Austin Jacksonville Seattle
	Knowing or helping neighbors	Austin Seattle
	Percentage reporting trust in city leaders/government	Austin Jacksonville Seattle
Satisfaction	Perceived quality of life	Seattle
	Percentage perceiving racism as a local problem	Jacksonville

[a] Indicators for Austin, Texas, and Jacksonville, Florida, can be found compiled in Greenwood, undated. That document includes some indicators for Seattle, Washington that are also included in Sustainable Seattle, 2004.

[b] Indicators for Seattle can be found in Sustainable Seattle, 2004.

[c] The Pikes Peak Sustainability Indicators Project is a partnership between Fort Carson and the governments, businesses, and citizens of the region that hosts Fort Carson's primary operations (El Paso, Fremont, and Pueblo counties). All of the Pikes Peak indicators in this table can be found in Pikes Peak Sustainability Indicator Project (PPSIP), 2005.

[d] This example is a summary of a detailed list of metrics (PPSIP, 2005).

Sixth, the Army needs to address policies and regulations that impede sustainability implementation. As discussed earlier, our interviews with installation staff identified several examples of Army policies, such as some of the USACE building design guidelines, that can stop or slow down sustainability implementation. The Army needs to create ways for installation staff that have encountered such policy limitations to bring them to the attention of Army policymakers so they can examine whether these policies or regulations should be modified.

Seventh, the Army needs to ensure that the true costs and benefits of sustainability projects are measured in terms of the triple bottom line, (i.e., mission, community, and environmental issues). In 2007, installations were not doing enough analyses to truly measure the full costs and benefits of sustainability projects to the installation and community. There are anecdotal examples, but they usually do not take into account the full range of costs and benefits. Analytical studies are needed to show the true savings and costs both in dollars and in impacts to mission, QOL, and the environment. Life-cycle costing is one of numerous approaches to apply. However, such assessments can be expensive and require specialized analytical and technical skills. It is difficult for installations do such assessments given the cost and skills required. However, such studies are needed so that the benefits and costs can be clearly documented and then used by other installations to help plan and implement sustainability projects. Common assessment methodologies need to be developed so that multiple installations can use them. Such analyses and methods are also needed to measure the success of sustainability activities. Clear metrics linked to installation sustainability goals need to be developed and tracked over time.[11] In addition to measuring progress, such metrics are needed to communicate progress, to help assign responsibility and accountability for sustainability results, and to help motivate behavior.[12]

Eighth, the Army needs more documentation, analysis, and transfer of lessons learned and success stories to help installations more effectively and efficiently implement sustainability. Installations have made many accomplishments and gained knowledge about how to implement sustainability, but this information is not being documented. Diverse sustainability champions and pioneers at leading installations (such as master planning and range personnel at Fort Bragg, range and P2 personnel at Fort Hood, and environmental management staff at Forts Carson and Lewis) have learned what to do and what not to do to implement sustainability projects successfully, but this information exists mostly in their heads. Installations just starting sustainability programs could benefit from their experience and technical help, especially from the more mature and advanced installation sustainability programs. Currently, these lessons and successes are not being transferred to other parts of the Army, other than through informal personal contacts and information sharing. There is a need for more standardized information sharing across installations about how to implement sustainability. There is a need for consistent and continually updated ISP documentation

---

[11] As noted earlier, later ISP planning has culminated in "action planning" sessions that included metrics to track progress against the installation's goals.

[12] For an example of how six different companies used pollution prevention metrics for such purposes, see The Business Roundtable, November 1993. For a good reference on metrics for sustainability in industry, see Global Environmental Management Initiative (GEMI), March 2007. For an overview of the importance of environmental assessments, accounting, and metrics (both qualitative and quantitative metrics) in environmental management, see Lachman et al., 2001, chapter 5.

across installations. More technical support is also needed for installations to implement an ISP.

Ninth, there needs to be a mindset and cultural change throughout the Army to truly integrate sustainability into installation operations. To achieve this change, it is important to provide education on sustainability throughout the Army, to all installation and unit staff. It is especially important to educate senior installation staff, such as garrison commanders and directorate leaders. Installations need classes for staff, soldiers, and family members that focus on sustainability. Installations that had more accomplishments found that such education helped their sustainability program. For example, Fort Hood includes sustainability in its environmental education classes, which has helped educate and motivate soldiers and installation staff to participate in the installation recycling and reuse programs.[13]

---

[13] The environmental management and sustainability literature also stresses the importance of education and training. For a good overview of how industry integrates training about environmental issues throughout an organization, see Lachman et al., 2001, pp. 143–148.

CHAPTER FIVE

Recommendations to Improve the ISP Development and Implementation Process

In this chapter we present our recommendations for improving the ISP development and implementation processes.

Recommendations for the ISP Development Process

Our recommendations for improving the ISP development process fall into two categories: revising the multi-workshop development process and providing HQDA guidance about issues to include and required documentation for ISPs. It is important to note that some of these recommendations may also affect ISP implementation, since some activities that support the development of an ISP can also help with implementation.

Streamline the ISP Development Process

As we discussed in the previous chapter, the multi-workshop process for developing ISPs that was being used through early FY07 needed to be streamlined. Our recommendation is that HQDA support one installation sustainability workshop to educate and motivate staff about sustainability, show commander support, and develop initial sustainability teams. One workshop should be sufficient to create initial support for the process and provide the other main benefits of the workshop process. It is also important that this workshop be customized to fit the installation's needs and to focus on its mission and QOL, as well as its environmental concerns. For example, educational information about sustainability successes should be carefully tailored for an installation based on its unique circumstances and interests. Since this research was conducted in 2007, ACSIM has streamlined the workshop process as recommended here, by holding fewer workshops and customizing them more to local installation needs.

Then, to develop its ISP, an installation needs technical support and assistance based on the knowledge and expertise of other Army installation sustainability experts who have successfully developed and implemented an ISP. People who have developed and helped implement an ISP at another installation should assist installations that are just starting to develop an ISP, with HQDA support. There are a number of possible approaches, including having installation staff who are developing an ISP visit installations that have already effectively developed and implemented an ISP; having knowledgeable IMCOM and installation staff experienced with ISP implementation, such as Fort Bragg and Fort Hood range control staff, visit their counterparts at the beginner installation;[1] having a monthly conference call among installations that are developing and implementing ISPs; and having Army sustainability experts assist the teams that are developing and writing sections of the ISP after the workshop. We think each of these methods should be considered where feasible and reasonable, given the unique circumstances and needs of each installation developing an ISP. An HQDA-level sustainability fund should pay for appropriate travel costs and ideally for the sustainability experts' time spent on providing assistance.[2] For example, if staff from an installation just starting the process visit Fort Lewis to learn about its sustainability program, this fund should cover the travel costs and Fort Lewis's staff time for meeting with the other installation's staff.

Provide HQDA Guidance for Developing ISPs

HQDA should provide guidance about issues to include in the ISP development process and establish requirements for ISP documentation. This guidance should include a list of installation functional staff to include in the process. This list should be designed to be comprehensive and to include all relevant functional staff on an installation. Such functional staff include DPW, the contracting office, PAIO, MWR, chaplain, master planning, DOIM, natural resources, DPTM, DRM, procurement, supply, health, family support services, and education staff. Installations should not be required to have all these organizations represented at the ISP development workshop, but they should at least be invited and kept informed about the process. However, the guidance should stress the need to ensure QOL staff participation, such as MWR, health, education, and family support services staff, given the lack of focus on such issues in ISP development and implementation. It should also stress the need to include operational units and other mission-related staff in addition to range staff.

This guidance should also provide a list of similar functional areas to consider as part of the long-term sustainability process, to help eliminate gaps. Such functional areas should include training and other mission functions; facilities and buildings development, management, and maintenance; energy, water, and transportation infra-

---

1 It is often less efficient to have installation staff from an experienced installation visit the installation starting an ISP. However, we found a couple of instances where this was helpful. Usually, it is more effective to have the staff starting an ISP visit the more experienced staff, where they can see the completed projects first hand.

2 HQDA paying for small amounts of installation staff time often is not feasible within the Army. However, HQDA could fund a sustainability position at an installation part time, such as 30 or 50 percent time over a year, assuming that that amount of time would be spent providing sustainability assistance to other installations, or alternatively it could provide some limited sustainability consulting help to assist at the installation in exchange for installation sustainability staff helping other visiting installation staff.

structure; products and materials management; biodiversity, species, ecosystem, and other natural resources management; community and regional interaction, especially related to land use, resource use, and ecosystem concerns; and QOL areas. QOL areas should include education; youth involvement; health and wellness; housing, work, and community facilities; recreation and culture; safety; family support; and overall MWR considerations.

Such guidance is designed to ensure that the ISP team structure, goals, and objectives consider these issues, especially those identified as gaps in current ISP activities, such as QOL, regional collaboration and outreach, conservation, and nonrange mission areas. Installations may choose the initial focus for ISPs based on local circumstances, but they should know that other key areas should be addressed over the long term by the ISP process. For example, an installation may not initially gain much participation in the sustainability process from MWR or health staff, but the sustainability program staff should continue to keep them informed and try to draw them into the process as it becomes more established.

This guidance should include standard reporting requirements, i.e., that each installation submit a baseline ISP and what should be in it. Such a baseline document should include current sustainability challenges, baseline sustainability conditions and accomplishments, sustainability goals, initial objectives, team organization, responsibilities, activities, targets, and metrics for measuring progress toward the goals.

In addition, the guidance should require that each installation submit brief annual progress reports describing the status of their ISP implementation. These reports should be designed not to overburden the installations with reporting requirements. Such yearly reports should include a brief summary of each ISP team's objectives; accomplishments during the past year toward those objectives (including tracking quantifiable performance metrics over time); future planned objectives, activities, and tasks; and challenges to meeting sustainability goals and objectives. Fort Carson and Fort Lewis have published yearly sustainability reports that are good models for such a requirement.[3]

Last, like other installation planning documents, the guidance should require installations to update ISPs every five years, including revisiting goals, objectives, and team structure.

Recommendations for the ISP Implementation Process

Next, we discuss recommendations to improve the ISP implementation process. These recommendations include HQDA guidance for more effective ISP implementation and support needed from other Army organizations to help with ISP implementa-

---

[3] See, for example, Fort Carson, 2006, and Fort Lewis, 2006.

tion, because organizations external to installations and their chains of command affect installation operations. The proposed recommendations will require resources and action, but these are needed investments if the Army wants ISP development and implementation to be successful across the Army.

Provide HQDA Guidance for Implementing ISPs

HQDA should provide guidance for implementing ISPs. This guidance should address some short-term issues that are easier to resolve as well as longer-term objectives. Such guidance should address nine key topics.

First, the Army should direct garrison commanders and other senior installation staff to support sustainability, including investing staff, funding, and other resources. To ensure support, sustainability should be included in their performance evaluations and their education. Key installation staff appraisals and performance reviews should include a sustainability element. Installation-level accountability for sustainability is needed, not just for sustainability and environmental staff, but for other relevant staff, such as master planning, range staff, etc. The Army should also include sustainability in the garrison commander's course, directorate education, and other education programs. All installation staff should learn about sustainability as part of their ongoing training and education.

Second, the Army should provide funding for installation sustainability staff. Experience suggests that at least 50 percent staff time is needed for a sustainability coordinator, particularly during the first few years of implementation, to grow and integrate the sustainability program throughout an installation. A sustainability coordinator is needed to motivate and educate staff throughout the installation about sustainability.

Third, the Army should address problems with project funding and return on investment. The Army should analyze and identify ways to solve the problems with different pots of money, lack of investment in sustainability because benefits are sometimes nonfinancial, etc. Installation and facility costing models may need to be adjusted to accommodate higher upfront costs of some products and services that result in long-term savings, or funding could be set aside for sustainability investments and allocated based on proposals from installations using life-cycle cost analysis. The Army should also examine mechanisms that would allow organizations that fund sustainability investments to benefit from the resulting savings and to reinvest savings from reduced energy, water, or disposal costs or the revenues generated from recycling programs. The Army should make sure that diverse installation organizations, not just DPW, range, and environmental programs, provide funding for sustainability projects.[4]

Fourth, HQDA should direct installations to integrate sustainability principles into key installation strategic planning, implementation, and operational documents,

[4] As noted earlier, since this research was conducted in 2007, ACSIM has been addressing some of these funding issues by including sustainability funding in the Army environmental requirements cost model.

such as the installation strategic and master plans, the INRMP, the IDG, ITAM guidance, and the range management plan. In December 2007 the updated AR 200-1 started to do this by directing installation commanders to incorporate sustainability principles into the installation strategic plan and other installation management plans. Over the long term, the ISP should be integrated with the installation strategic plan to integrate sustainability into key installation business processes. However, the integration of the ISP and the installation strategic plan must be done carefully to ensure that the sustainability focus and long-term goals are not lost or deemphasized.[5] Lean Six Sigma business process reengineering efforts should be used to improve sustainability efforts.

Fifth, the Army should ensure that installations place greater emphasis on the gaps that we identified in implementation, including QOL issues; regional collaboration and outreach; ecosystem management and other strategic natural resource management activities; and nonrange mission areas. Regional collaboration is especially needed in strategic areas such as growth management and where natural resource constraints are likely to arise in the future, such as water-quality concerns in North Carolina near Fort Bragg. HQDA should direct organizations such as AEPI and USACE to fund or conduct more strategic assessments to help identify strategic regional priorities for sustainability implementation. The CERL SSA pilot study, which helped identify the regional water-quality trends and concerns around Fort Bragg, is a good example of the type of analysis that could help installations identify regional sustainability needs and priorities.

Sixth, the Army should create a process to identify and address Army policy and funding issues that impede sustainability implementation. Installations that identify a policy roadblock, such as an Army regulation, should be able to communicate that sustainability impediment to the appropriate organization. Some centralized coordination may be needed to ensure that the policy is changed or a process is put into place, such as a clear and easy exemption process, that removes the sustainability impediment. Funding issues are more complicated, but again an installation should be able to communicate such problems to HQDA so that mechanisms can be developed to address them.

Seventh, HQDA should ensure that the true costs and benefits of sustainability projects are measured in terms of the triple bottom line. Since it is difficult for installations do such assessments given the cost and skills required, Army headquarters and regional organizations, such as ACSIM and IMCOM, could provide resources to con-

5 As was discussed earlier, having a separate ISP process to start with has been beneficial because it helps to educate and motivate staff to begin implementing sustainability and develop sustainability goals, especially since implementation staff are actively involved in the ISP development and implementation process. If sustainability principles are just integrated directly into the installation strategic plan there may not be as much motivation, interest, or knowledge to implement sustainability, and there is a danger that sustainability is no longer focused on as much in the plan and its implementation.

duct some initial assessments and document them so their value can be measured and lessons can be transferred. Such studies should also develop metrics and methodologies that could be applied by other installations, to make it easier for installation staff to conduct such assessments on their own. The Army could also provide a standing small consulting-style group to help perform this task upon request.

Eighth, the Army should share sustainability lessons learned across installations. HQDA should provide support for documenting and transferring lessons learned. It also should ensure that this information, including all installation ISPs and yearly status reports and any sustainability project assessments and methodologies, are available at a central sustainability web site. There are a number of other mechanisms that could be used for information sharing. The mechanisms recommended above to help installations develop ISPs (such as having sustainability staff visit an installation with an established ISP program and monthly conference calls) should be employed to help with sustainability implementation as well. Programs for Army technical sustainability experts to visit other installations and help them develop and implement a new ISP could focus on selected topics or staff. For example, a range expert from Fort Hood or Fort Bragg could help range staff at a similar installation that was just starting its sustainability process. In addition, ACSIM and IMCOM should sponsor both regional and national ISP conferences. An HQDA-level sustainability fund should help pay for installation staff to attend and ensure that leading installations share information. These conferences could also include sustainability training workshops on key topics, such as LEED implementation and range sustainability.

Ninth, HQDA should be helping to create a cultural change with its sustainability guidance. The Army should ensure that sustainability is viewed as a long-term, strategic, Army-wide program, not just an environmental one. The 2007 "Strategic Plan for Army Sustainability,"[6] which tried to integrate sustainability into all parts of the Army, was a good first step at potential guidance language, but this document remains in draft form. The Army should issue some guidance like what was in this draft document and ensure that it is implemented throughout the Army. Sustainability should be seen as being as important as safety. The way the Army integrated safety concerns throughout its installations is a good model.

Other Army Organizations Also Need to Help with ISP Implementation

Many organizations that are not currently or directly involved in installation sustainability need to help with ISP implementation.

To address the education and training needs, HQDA G-3/5/7 and TRADOC should include sustainability training in garrison and unit commander staff education.

---

6 U.S. Army, 2007(c).

The Army should integrate sustainability into Army-wide guidance and standards that affect installation operations, including USACE standards of excellence for building and facility designs and range design guides (such as *Training Circular 25-8: Guidance for Ranges*), ITAM guidance, and budgeting, programming, and procurement guidelines. BRAC transformation and implementation guidance also should mention the importance of and the need to follow installation sustainability practices, especially implementing LEED standards and addressing QOL considerations. HQDA guidance on ACUB and INRMPs should also include sustainability.

Army headquarters and command organizations should provide more support for sustainability, including G-3/5/7, USACE, and nonenvironmental functional organizations within IMCOM and ACSIM. Such support should ensure that their guidance, policies, practices, and training includes sustainability elements. It can also include providing funding for staff to participate in sustainability programs or to implement sustainability projects. For example, USACE can change their building design standards to address the need to accommodate sustainability features, provide more education to construction and contracting staff about the need to ensure that LEED is implemented in new construction projects, and provide more funding for sustainability R&D studies.

Conclusions

Installations have accomplished a great deal in developing and implementing ISPs despite limited resources and guidance. Given such accomplishments and the experiences of sustainable community and industry activities, successful ISP implementation has the potential to significantly benefit Army missions, QOL, and the environment. However, to promote the successful development and implementation of ISPs across the Army, the Army should provide sufficient resources and technical support; ensure that installations fully support ISP development and implementation; and provide HQDA guidance for the development and implementation of ISPs. The Army should also assess, document, and transfer sustainability lessons learned; better collect and address Army policy issues that impede sustainability implementation; and provide appropriate funding mechanisms to support installation sustainability activities.

APPENDIX A

Background on LEED and Army Green Building Standards

This appendix provides background information about the history, development, and implementation of "green" building standards in the United States and in the United States Army.

U.S. Green Building Council (USGBC)

The national green building coalition, known as the U.S. Green Building Council (USGBC), was formed in the mid-1980s and was officially incorporated in 1993 as a nonprofit organization. The founders of the USGBC consisted of professionals, leaders in industry, and federal agencies such as the Department of Energy and the Naval Facilities Engineering Command (NAVFAC). Today, the USGBC consists of more than 12,000 organizations that are trying to advance structures that are environmentally responsible, profitable, and healthy places to live and work. Members include building owners and end-users, real estate developers, facility managers, architects, designers, engineers, general contractors, subcontractors, product and building system manufacturers, government agencies, and nonprofit organizations.

The USGBC is dedicated to sustainable building design and construction. Its mission is to "transform the way buildings and communities are designed, built, and operated, enabling an environmentally and socially responsible, healthy, and prosperous environment that improves the quality of life."[1] The USGBC provides green building resources, education, and information to students and professionals. The USGBC is most widely known for the development of the LEED (Leadership in Energy and Environmental Design) building standard and the certification process for LEED-rated buildings. LEED is a voluntary, consensus-based national rating system for developing high-performance, sustainable buildings.

Green buildings are becoming more and more common throughout the United States and the world, and many follow the LEED standards and have benefited from the USGBC's educational and informational programs. There are now more than 1,000

[1] U.S. Green Building Council, 2007(a).

LEED-certified and over 7,300 LEED-registered buildings in the United States. In fact, LEED has become the leading green building rating system in the United States. In addition to all 50 U.S. states, 41 countries are home to a LEED building.[2] With global climate change awareness increasing at a rapid rate and an unprecedented public interest in sustainability, many countries have begun to develop organizations similar to the USGBC or simply to adopt USGBC policy with specific changes to meet their countries' needs. For example, Canada's Green Building Council (CGBC), formed in 2002, adapted the U.S. LEED rating system to fit its colder climate, and "more than 230 construction projects across Canada now carry a LEED designation."[3]

Leadership in Energy and Enviornmental Design Rating System

LEED was designed as a practical rating tool for green building design and construction certification that provides measurable results for building owners and occupants. This green building standard was also designed to encourage and accelerate global adoption of sustainable green building and development practices through the implementation of universally understood and accepted tools and performance criteria. LEED addresses all building types and emphasizes state-of-the-art strategies for sustainable site development, water savings, energy efficiency, materials and resources selection, and indoor environmental quality. Buildings can be LEED-certified, which means that a qualified third party has validated a project's green features and verified that the building is operating exactly the way it was designed to.

The first iteration of LEED (version 1.0) was released in 1998. The USGBC recognized the need for continuous improvement and has been updating the rating tool in order to keep up with the emerging green market of products, suppliers, and best practices. The current LEED standard (version 2.2), originally published in October 2005, now includes amended updates that were incorporated as recently as June 2007. Version 2.2 is the current required standard for newly constructed buildings, i.e., LEED for new construction (LEED-NC) projects seeking certification. The LEED-NC rating system is designed to guide and distinguish high-performance, more sustainable commercial and institutional projects, including office buildings, multi-unit residential buildings, manufacturing plants, and laboratories.

LEED Point Rating System

LEED is based on a simple point rating system; more points indicate a higher level of sustainability achievements. Namely, building projects earn LEED points for satisfying specific green building criteria. Within each of the six LEED credit categories, build-

[2] U.S. Green Building Council, 2007(a).

[3] Hadekel, April 6, 2007.

ing projects must satisfy particular prerequisites and earn points. The six categories are sustainable sites, water efficiency, energy and atmosphere, materials and resources, indoor environmental quality, and innovation in design: projects can earn extra points for green building innovations. The number of points the project earns determines the level of LEED certification the project receives. Based on a scale of 0–69, points are totaled to achieve one of four progressive levels of certification: Certified, Silver, Gold, and Platinum. The LEED portfolio of rating tools has been refined to include specific types of buildings: commercial interiors, schools, multiple buildings with one owner (such as campus settings and military installations), homes, health care, and existing buildings for continuing operations.

Table A.1 contains a checklist that shows the categories and total points required under LEED version 2.2 for new construction. To earn a LEED certification, a project must satisfy all prerequisites and the minimum number of points outlined in this table. For example, conducting pollution prevention during site construction is a required prerequisite, while there is a one-point credit for brownfield redevelopment. Some categories have both prerequisite conditions and minimum point requirements as part of the total for each LEED rating.[4]

LEED Certification Process

The USGBC offers online assistance for building owners preparing and applying for certification; however, the required documentation is time consuming and must occur continuously throughout the entire design and construction process in order to be successful. Registration for a building is nominal, $450 for USGBC members and $600 for nonmembers. However, final certification fees can vary depending on square footage and certification level desired. Certification fees are approximately $0.035 for design and construction review per square foot for a building of 50,000 to 500,000 square feet. For anything less than 50,000, there is a fixed fee of $1,750, and for buildings greater than 500,000, the fixed fee is $17,500. For example, in 2004, for a Silver-certified 49,500-square-foot, $11.5 million Consolidated Support Facility (CSF) at Edwards Air Force Base (AFB), certification cost $1,750. The certification fee pays for the USGBC to review the extensive paperwork that documents the building's qualifications for the LEED standards. However, the real costs to "green" a building are associated with selection of green materials and incorporating sustainable features and preparing the certification documentation that must go to the USGBC. This documentation must include extensive receipts to verify that green building features, such as environmentally friendly light bulbs, were purchased and installed. Often it requires a dedicated person throughout the building process to collect the paperwork and then submit it to the USGBC. This can add another 1 to 2 percent to the overall building construction cost. For the Edwards AFB CSF, it was around 1 percent or about $115,000.

---

[4] For more details on these requirements, see U.S. Green Building Council, October 2005.

Table A.1
LEED for New Construction (Version 2.2) Project Checklist

Category	Activities	Points Earned or If Required
Sustainable sites		*14 for the category*
	Construction activity pollution prevention	Required
	Site selection	1
	Development density and community connectivity	1
	Brownfield redevelopment	1
	Alternative transportation: public transportation access	1
	Alternative transportation: bicycle storage and changing rooms	1
	Alternative transportation: low-emitting and fuel-efficient vehicles	1
	Alternative transportation: parking capacity	1
	Site development: protect or restore habitat	1
	Site development: maximize open space	1
	Stormwater design: quantity control	1
	Stormwater design: quality control	1
	Heat island effect: non-roof	1
	Heat island effect: roof	1
	Light pollution reduction	1
Water efficiency		*5 for the category*
	Water efficient landscaping: reduce by 50%	1
	Water efficient landscaping: no potable use or no irrigation	1
	Innovative wastewater technologies	1
	Water use reduction: 20% reduction	1
	Water use reduction: 30% reduction	1
Energy and atmosphere		*17 for the category*
	Fundamental commissioning of the building energy systems	Required
	Minimum energy performance	Required
	Fundamental refrigerant management	Required
	Optimize energy performance (OEP):	1 to 10
	OEP: 10.5% new buildings (NB) or 3.5% existing building (EB) renovations	1
	OEP: 14% NB or 7% EB renovations	2
	OEP: 17.5% NB or 10.5% EB renovations	3
	OEP: 21% NB or 14% EB renovations	4
	OEP: 24.5% NB or 17.5% EB renovations	5
	OEP: 28% NB or 21% EB renovations	6
	OEP: 31.5% NB or 24.5% EB renovations	7
	OEP: 35% NB or 28% EB renovations	8
	OEP: 38.5% NB or 31.5% EB renovations	9
	OEP: 42% NB or 35% EB renovations	10
	On-site renewable energy	1 to 3
	On-site renewable energy: 2.5%	1
	On-site renewable energy: 7.5%	2
	On-site renewable energy: 12.5%	3

Table A.1 (continued)

Category	Activities	Points Earned or If Required
Energy and atmosphere	Enhanced commissioning	1
	Enhanced refrigerant management	1
	Measurement and verification	1
	Green power	1
Materials and resources		*13 for the category*
	Storage and collection of recyclables	Required
	Building reuse: maintain 75% of existing walls, floors, and roof	1
	Building reuse: maintain 100% of existing walls, floors, and roof	1
	Building reuse: maintain 50% of interior nonstructural elements	1
	Construction waste management: divert 50% from disposal	1
	Construction waste management: divert 75% from disposal	1
	Materials reuse: 5%	1
	Materials reuse: 10%	1
	Recycled content: 10%	1
	Recycled content: 20%	1
	Regional materials: 10% extracted, processed, and manufactured regionally	1
	Regional materials: 20% extracted, processed, and manufactured regionally	1
	Rapidly renewable materials	1
	Certified wood	1
Indoor environmental quality		*15 for the category*
	Minimum indoor air quality (IAQ) performance	Required
	Environmental tobacco smoke control	Required
	Outdoor air delivery monitoring	1
	Increased ventilation	1
	Construction IAQ management plan: during construction	1
	Construction IAQ management plan: before occupancy	1
	Low-emitting materials: adhesives and sealants	1
	Low-emitting materials: paintings and coatings	1
	Low-emitting materials: carpet systems	1
	Low-emitting materials: composite wood and agrifiber products	1
	Indoor chemical and pollutant source control	1
	Controllability of systems: lighting	1
	Controllability of systems: thermal control	1
	Thermal comfort: design	1
	Thermal comfort: verification	1
	Daylight and views: daylight 75% of spaces	1
	Daylight and views: views for 90% of spaces	1

Table A.1 (continued)

Category	Activities	Points Earned or If Required
Innovation and design process		*5 for the category*
	Innovation in design: provide specific title (up to four different ones)	1 for each title
	LEED-accredited professional	1

NOTE: Total points required for each certification type: 26–32, Certified; 33–38, Silver; 39–51, Gold; 52–69, Platinum.

SOURCE: U.S. Green Building Council, October 2005.

Adding the actual physical equipment to making the building greener, such as solar panels and a more energy-efficient HVAC, can increase initial construction cost another 2 to 7 percent. For the Edwards AFB CSF it was about 3 percent, which added another $345,000 to the overall project cost. The benefit here is in increased energy savings that result in paybacks down the road. In sum, making a building greener costs 2 to 7 percent more of the total building cost, and to certify it to the LEED standard can add another 1 to 2 percent. Generally, higher costs are associated with achieving higher LEED certifications, such as a Platinum rating; however, reduced costs can be achieved through adequate preconstruction planning and enlisting team members knowledgeable on green building practices.

LEED for Existing Buildings

LEED for existing buildings (LEED-EB) was developed to take certified green buildings to the next step of staying green for life. The LEED-EB rating system is designed to help building owners and operators measure operations, improvements, and maintenance on a consistent scale, with the goal of maximizing operational efficiency while minimizing environmental impacts. It addresses whole-building cleaning and maintenance issues (including chemical use), recycling programs, exterior maintenance programs, and systems upgrades. LEED-EB can be applied both to existing buildings seeking LEED certification for the first time and to projects previously certified under LEED for new construction.[5] The USGBC continues to develop this rating system due to increased concern that one-time certification of a building does not ensure that it operates in a more sustainable way and that the entire lifespan of a building should be examined in order to maintain a green certificate of achievement.

[5] U.S. Green Building Council, 2007(b).

U.S. Army Sustainable Project Rating Tool Standard

In 2001, the Army developed its own green building standard, the Sustainable Project Rating Tool (SPiRiT) which is based on a point system similar to LEED. As the largest federal building owner, ACSIM staff realized there was an increasing need to reduce energy usage. It would be necessary to take associated costs along the entire life cycle of buildings into consideration. ACSIM staff asked the USACE Construction CERL in Champaign, Illinois to tackle the problem. After comparing LEED and other commercial rating tools, they realized the Army needed something different and created SPiRiT. "Unlike LEED [version 1.0], SPiRiT includes O&M (Operations and Maintenance) issues and flexibility in design to allow for building modifications as needs change."[6]

USACE has a license agreement with the USGBC permitting the Army to use the LEED name as part of SPiRiT.[7] In 2001, after the Army reached an agreement with the USGBC, SPiRiT was promoted as the tool for rating sustainable design and development (SDD) and "green" buildings on Army installations. Although LEED 2.0 was the base, changes were made in language, references to Army publications were added, and the site criteria were adapted to better fit Army installation concerns with multiple buildings and O&M costs. ACSIM required that beginning May 1, 2001, all installations must attempt to achieve a minimum of a SPiRiT Bronze rating for new construction. Later in 2003, a follow-up memorandum was issued increasing this requirement to SPiRiT Gold.

LEED Becomes the Army Green Building Standard

In order to divest itself from the responsibility of continuously updating and maintaining a separate rating tool used solely by Army contractors, CERL continued its collaboration with the USGBC with the intent to convert to the LEED rating system as it became more appropriate for military installations. In 2006, USACE published guidelines for using LEED instead of SPiRiT:

> Starting with the FY06 programs, all vertical projects with climate-controlled buildings are required to achieve a SPiRiT Gold rating. Renovation, upgrade and rehabilitation projects are also required to reach a Gold rating unless the cost increase will put the project over the "50% of the replacement" cost threshold. In this case, a waiver from ACSIM is required.[8]

---

[6] Cassidy, November 2003, pp. 1–47.

[7] Beranek, May 1, 2001.

[8] U.S. Army Corps of Engineers (USACE), November 7, 2003 (revised 10 May 2006).

Starting with the FY08 Military Construction, Army (MCA) program, projects with climate-controlling abilities are required to be capable of achieving USGBC certification at the LEED-NC Silver rating.[9] The LEED-NC standard is now officially the Army standard. The Army is also concerned about O&M costs and building performance throughout the lifespan of its facilities, so it has an existing building requirement. FY07 policy requires a minimum certification level for LEED-EB on renovation work exceeding $7.5 million. However, this policy is under revision. The Army is considering removing the LEED-EB requirement for major retrofit projects.[10]

Because of the additional costs of certification (usually not projected years earlier), the Army did not require any of its LEED projects to be officially certified by the USGBC in 2007, only that projects be built to "certifiable" standards. However, this policy was also under review by ACSIM and USACE late in 2007. The Army is considering using project funds to pay for full-fledged USGBC certification. Many installations (both Army and Air Force), as well as general contractors, are supportive of actual certification, because with it comes prestige and validation. The Director of Public Works at Fort Bragg recently said that "the positive publicity of getting official certification may build momentum to further energy efficient, green construction," and that is worth the extra money.[11]

---

9 Besham, 2006.

10 U.S. Army Corps of Engineers (USACE), October 12, 2007.

11 Rawlins, November 23, 2007.

APPENDIX B

Quality of Life Issues in Sustainability Activities

The concept of "quality of life" is reflected in the community aspect of the Army's triple bottom line of sustainability.[1] What is meant by community and QOL? This appendix describes definitions and concepts of QOL in sustainability planning and implementation as they are described in the available literature and approached by communities that are leaders in sustainability planning. These communities include QOL in the social aspect of their triple bottom line of sustainability. This appendix also presents examples of metrics that sustainability leaders are using to measure and assess progress in QOL issues.

Quality of Life Issues Are Inconsistently Defined

A number of communities in the United States have undertaken ambitious sustainability programs. As their programs have evolved, each has seemed to struggle with a central question: What does "sustainability" mean for an individual citizen and for the social environment? In other words, how should sustainability managers and proponents measure the effects of their programs on individuals' everyday well-being? Herein lies the concept of "quality of life." Several communities have sought to measure this quality in ways that could be informative to similar Army efforts.[2]

Early attempts to quantify QOL issues focused on measures that were readily available: economic metrics. The idea was that increased income and productivity were useful proxies for quality of life. At a broad level, measures such as gross domestic product (GDP) were assumed to be useful measures of social well-being and standard of living, even though the developers of the GDP concept warned against this conceptual extension. It is becoming more apparent that such purely economic metrics are insufficient to measure QOL.[3] Several organizations have attempted to define metrics spe-

---

1 As discussed earlier, the "triple bottom line of sustainability" refers to mission, community, and environmental issues.

2 For example, see Mueller, 1999, and Greenwood, 2001.

3 Greenwood, undated.

cifically for assessing progress in QOL issues, and these are summarized below. First, however, one should consider what is encompassed by QOL issues in sustainability.

What Is Meant by Quality of Life?

QOL issues in sustainability have been described in many different ways. Other sustainability concepts are more easily defined, particularly those that have grown from compliance-based management systems, such as pollution prevention. Because nearly every aspect of installation management affects the people who work and live on or near the installation, it could be argued that every decision about installation management affects QOL. Reducing air emissions on an installation may improve environmental compliance, but it also can affect, for example, local health and well-being: a QOL. In many cases, sustainability initiatives that are designed to address other objectives are also QOL issues; for example, employment and education rates may indicate economic well-being but obviously have significant implications for QOL as well.

Metrics used to assess QOL will vary by community, but interesting similarities can be found among sustainable community activities. Community initiatives provide some definitions and objective statements of sustainability that can be useful in defining and measuring QOL. We present three community examples here. The community of Jacksonville, Florida, defines QOL as "a feeling of well-being, fulfillment, or satisfaction resulting from factors in the external environment."[4] The sustainable community program of Seattle, Washington, asks: "How do we make difficult tradeoffs and balanced judgments that take everyone's interests into account, including those of our children and grandchildren?"[5] The community of Austin, Texas, combines these views of well-being and concern for future generations: their goal includes "recognizing the interdependence of the environment, economic development and social equity . . . with a decision-making climate that invests in what is good for today without compromising the future for our children, a climate that benefits each person, and the common good."[6] Although these comprehensive definitions do not imply any specific metrics, they show how QOL issues pervade all aspects of sustainability planning.

The available literature is replete with philosophical discussions of what QOL means and what factors affect it, including discussions of a "sense of place,"[7] proposed indicators of social capital and economic development,[8] and more overarching discus-

---

4 Jacksonville Council, 2001.

5 Sustainable Seattle, 2004.

6 Austin, Texas, 2000, and Greenwood, undated.

7 Norton, 1991 and 2005.

8 Putnam, 1993.

sions of how QOL might be integrated with such concepts as sustainability, social environments, and wealth.[9] However, Army sustainability programs need more clearly defined concepts and measures of QOL in order to develop initiatives to address the "community" aspect of the "triple bottom line" in ISP implementation. Therefore, next we provide some examples of how communities that have implemented sustainability programs have addressed QOL issues—and the indicators they have developed to measure their progress.

Examples of Quality of Life Metrics

Several communities have established sustainability initiatives that address QOL issues. As mentioned above, many initiatives that target other sustainability areas, such as the environment, energy, or the economy, also affect QOL, but some initiatives are specifically directed at improving QOL. Because so many factors affect societal and personal well-being, it is not possible to compile an exhaustive list of what should be measured when assessing progress in improving QOL. Our review of the community sustainability literature found that the issues measured are disparate and variable among communities, as might be expected if one considers that sustainability is inherently a local initiative. In fact, it has been pointed out that QOL issues vary by individual, as "each person has a unique utility function."[10] Several authors, agencies, and communities have suggested various metrics or indicators that could be used to measure QOL issues.[11] However, there are some common subject areas that QOL metrics typically address:

- Economy, wealth, and employment
- Housing and community facilities
- Land use and density
- Education and youth involvement
- Health and wellness
- Communications (e.g., internet access)
- Recreation and culture
- Crime and safety
- Civic involvement and governance
- Spiritual opportunity/participation
- Less tangible measures, such as "fulfillment" and "satisfaction"

---

[9] Phillips, 2006, and Schafer, Nolting, et al., 2004.

[10] Flynn, Berry, et al., 2002.

[11] For examples, see Estes, 1997, Andre and Bitondo, 2001, United Nations, 2004, and Shackman, Liu, et al., 2005.

To illustrate the types of indicators in each of these areas we present some examples in Table B.1 from Jacksonville, Florida; Seattle, Washington; Austin, Texas; and the Pikes Peak Sustainability Indicators Project (PPSIP). Jacksonville was one of the earliest community sustainability programs to develop QOL indicators in 1986. Seattle followed in the 1990s, and Central Texas, focused around the community of Austin, published a set of indicators in 2000.[12]

Table B.1
Examples of QOL Sustainability Indicators

Category	Indicator	Community
Economy, wealth, and employment	Percent of population in poverty	Pikes Peak[a]
	Child poverty and overall poverty rate	Austin[b] Seattle[c]
	Percentage of children in families below basic need level or on school lunch program	Jacksonville Seattle
	Rate of change of median income	Austin
	Share of income by income group (income distribution)	Seattle
	Monthly hours worked for basic needs	Seattle
	Livability/affordability index: percent of people earning below livable compensation and percent of new regional jobs providing livable compensation, percent of income of people earning at and/or below regional median devoted to necessities	Pikes Peak
	Percentage of new businesses surviving 3+ years	Austin
	Percent of new jobs filled by local residents	Pikes Peak
	Percent of total jobs and new jobs that are easily accessible to economically challenged neighborhoods	Pikes Peak
	Percentage of total jobs in private sector	Austin
	Percentage of jobs from top ten private employers	Austin Seattle
	Percentage of jobs in top ten industry sectors	Austin
	Unemployment rate	Pikes Peak
Housing and community facilities	Percentage of households able to purchase median-priced home	Austin
	Existing and affordable housing	Pikes Peak
	Rental affordability: Percent of households for which average rent would be less than 35% of household income	Austin
	Housing within proximity to commercial and transit services	Pikes Peak
	Family self-sufficiency index (including access to health insurance, cost-effective mobility, daycare)	Pikes Peak
	Percentage of licensed child care workers replaced annually	Austin
	Percentage of streets within 1,000 feet of urban villages, schools, and social services that have sidewalks	Seattle
	Streets with bike lanes	Seattle
	Direct air flight destinations	Jacksonville
	Number of community garden plots available	Seattle

[12] Sustainable Seattle, 2004, Austin, Texas, 2000, Jacksonville Council, 2000, and Greenwood, 2001.

Table B.1 (continued)

Category	Indicator	Community
Land use and density	People per acre in residential zones (i.e., bedrooms per acre of residential zoning)	Pikes Peak
	Average commute times	Pikes Peak
	Average work commute/vehicle miles	Austin Jacksonville Seattle
	Miles traveled per capita and annual fuel consumption	Seattle
	Annual population change rate	Seattle
Education and youth involvement	Number of students per classroom and per teacher	Pikes Peak
	Percent of all public education students meeting proficiency standards	Pikes Peak
	Annual dropout and graduation rates	Pikes Peak
	High school graduation rate	Jacksonville
	High school completion rate by cohort and ethnicity	Seattle
	Percentage of students at or above grade level	Austin Jacksonville
	Ethnic diversity of teachers compared to student diversity	Seattle
	Citizen volunteer hours per student in public schools	Seattle
	Satisfaction of parents with the educational system	Pikes Peak
	Percent of adults with degrees and/or in higher education	Pikes Peak
	Adult literacy	Seattle
	Students reporting local or national volunteer service	Seattle
Health and wellness	Health insurance coverage	Austin Jacksonville
	Health care expenditures per capita	Seattle
	Emergency room use for nonemergencies	Seattle
	Percent of school districts with mandatory health education	Pikes Peak
	Physical activity, smoking, obesity, drug use, sexually-transmitted disease, immunizations, and reduction of death due to preventable conditions[d]	Pikes Peak
	Percentage reporting good health status/health care	Austin Jacksonville
	Low birthweight rates	Seattle
	Infant mortality	Jacksonville
	Youth alcohol use	Jacksonville
	Cigarette sales	Jacksonville
	Lung cancer deaths	Jacksonville
	Suicides	Austin
	Good air quality days	Jacksonville
	Number of days not meeting national ozone standards	Austin
	Number of days air quality was good, moderate, unhealthful, very unhealthful or hazardous, according to Puget Sound Air Control Agency	Seattle
	Annual number of exceedences of National Ambient Air Quality Standards for carbon monoxide and coarse particulate matter	Seattle
	Childhood asthma hospitalizations	Seattle

Table B.1 (continued)

Category	Indicator	Community
Communications	Households with internet access	Pikes Peak
Recreation and culture	Tax funding supporting cultural and leisure activities	Pikes Peak
	Percent attending artistic or cultural activities	Austin Jacksonville
	Number of arts organizations and attendance	Seattle
	Number of students per art teacher in public schools	Seattle
	Library circulation	Jacksonville
	Library circulation and community center visits per capita	Seattle
	Open space/park acreage	Austin Jacksonville
	Areas within 1/8th mile of open spaces	Seattle
Crime and safety	Crime rate	Austin
	Juvenile crime	Seattle
	Reports of family violence or child abuse	Austin Jacksonville
	Vehicle accidents	Austin Jacksonville
	Pedestrian/vehicle injuries and fatalities	Seattle
	Resident perception of safety and crime statistics, including juvenile arrests	Pikes Peak
	Percentage who feel safe walking at night	Jacksonville
Civic involvement and governance	Volunteerism, charitable donations and individual satisfaction with community involvement	Pikes Peak
	Time spent in volunteerism	Austin Jacksonville
	Civic participation rates	Pikes Peak
	Percentage of registered voters voting in local elections	Austin Jacksonville Seattle
	Number of neighborhoods with neighborhood associations	Pikes Peak
	Knowing or helping neighbors	Austin Seattle
	Percentage reporting trust in city leaders/government	Austin Jacksonville Seattle
	Quality of elected officials and executives	Pikes Peak
Satisfaction	Perceived QOL	Seattle
	Percentage perceiving racism as a local problem	Jacksonville
	Racial disparities in juvenile courts	Austin Seattle

[a] The Pikes Peak Sustainability Indicator Project (PPSIP) is a partnership between Fort Carson and the governments, businesses, and citizens of the region that hosts Fort Carson's primary operations (El Paso, Fremont, and Pueblo counties). All of the Pikes Peak indicators in this table can be found in PPSIP, 2005.

[b] Indicators for Austin and Jacksonville can be found compiled in Greenwood (undated). That document includes some indicators for Seattle that are also found in Sustainable Seattle, 2004.

[c] Indicators for Seattle can be found in Sustainable Seattle, 2004.

[d] This example is a summary of a detailed list of metrics (PPSIP, 2005).

The PPSIP is particularly relevant, since it involves Fort Carson. The PPSIP is a partnership between Fort Carson and other federal agencies, local governments, businesses, and citizens of the region surrounding that garrison. The project was started in 2003 and published sustainability indicators in 2005.[13] The project identified a "universe" of indicators that could be monitored to determine sustainability progress, and it narrowed this list to "recommended" indicators that will be used to measure progress at Fort Carson and its surrounding community. Many of these indicators are related to QOL issues, and some are shown in Table B.1. Examples of these indicators with relevance to Army installations include: family self-sufficiency index (access to health insurance, cost-effective mobility, daycare, etc.); existing and affordable quality housing; housing within proximity to commercial and transit services; average commute times; number of students per classroom and per teacher; percent of students meeting proficiency standards; annual student dropout and graduation rates; satisfaction of parents with the educational system; resident perception of safety and crime statistics, including juvenile arrests; and physical activity, smoking, obesity, drug use, sexually-transmitted disease, immunizations, and reduction of death due to preventable conditions.

It should be noted that the Pikes Peak effort, like many others, has been criticized for lacking the readily available data necessary to measure progress in QOL issues.[14] Recognizing that refining metrics is an ongoing part of sustainability implementation, the PPSIP report notes that some indicators are better defined than others and that some of the indicators will be further refined in a second phase of their project.

It is also important to note that the United States is not the only country where communities have developed sustainability initiatives and QOL indicators. Many other countries have similar activities. One noteworthy example from a QOL perspective is in New Zealand. The eight largest cities in New Zealand embarked on a sustainability program in the 1990s, and "well-being and sustainability" initiatives were codified there by law in 2002. The law directs local governments to:

> Enable democratic local decision-making and action by, and on behalf of, communities; and promote the social, economic, environmental, and cultural well being of communities, in the present and for the future.[15]

In order to "assess sustainable quality of life," the New Zealand sustainability community has developed a set of QOL indicators. Some different examples of QOL indicators that are potentially relevant to Army installations, in six key categories, are shown in Table B.2. The table does not include all of the New Zealand indicators nor all their categories of indicators, but a sample that are of interest for the Army's ISP process.

---

[13] PPSIP, 2005.

[14] Greenwood, undated.

[15] New Zealand, 2003.

Table B.2
Examples of QOL Sustainability Indicators from New Zealand

Indicator Category	Sample Indicators
Built environment	Look and feel of the city, city green space, graffiti, noise pollution, traffic and transportation, and public transportation
Economic standard of living	Income, costs, household expenditure, and social deprivation
Health	Life expectancy, low birth weights, infant mortality, teenage parents, diseases, access to general practitioners, mental health, health status, and modifiable risk factors
Knowledge and skills	Early childhood education, school decile ratings, suspensions, qualification levels, and community education
Safety	Perceptions, child safety, road casualties, and crime
Social connectedness	Diversity, community strength and spirit, and electronic communication

Different communities also often use the same or similar QOL indicators. Table B.3 shows some examples of common and similar QOL indicators that were used by Seattle, Austin, and Jacksonville sustainability programs. It also lists a few indicators that were unique to one of these communities to show some of the diversity. It is important to note that not all of the QOL indicators in this table are directly relevant for installation sustainability planning.

In summary, there are many QOL indicators and approaches described in the literature on sustainable communities that could help Army installations better understand how to incorporate QOL issues in their sustainability programs.

It also is important to note that industry as well is addressing QOL concerns in its sustainability activities. For example, the GEMI Sustainable Development Planner includes work/life balance issues: "Promoting balance between work and personal or family life. Examples of programs and practices include flexible work options, dependent care services, time-off policies, or health and wellness programs."[16]

Table B.3
Comparing QOL Indicators Across Three Communities

Indicator	Seattle	Austin	Jacksonville
Child poverty or overall poverty rate	X	X	
Percentage of children in families below basic need level or on school lunch program	X		X
Median home price/median income			X
Percentage of households able to purchase median priced home		X	
Rental affordability[a]		X	
Percent average rent above affordability for low-income households	X		
Rate of change of median income		X	

[16] Global Environmental Management Initiative, 2006, p. 44.

Table B.3 (continued)

Indicator	Seattle	Austin	Jacksonville
Percentage of new businesses surviving 3+ years		X	
Percentage of total jobs in private sector		X	
Percentage of jobs from top ten private employers	X	X	
Percentage of jobs in top ten industry sectors		X	
Good air quality days	X		X
Number of days not meeting national ozone standards		X	
Open space/park acreage		X	X
Percentage living near urban open space	X		
Health insurance coverage		X	X
Emergency room use for nonemergencies	X		
Percentage reporting good health status/health care		X	X
Low birth weight rates	X		
Infant mortality			X
Youth alcohol use			X
Cigarette sales			X
Lung cancer deaths			X
Percentage feel safe walking at night			X
Suicides		X	
Crime rate		X	X
Reports of family violence or child abuse		X	X
Percentage of registered voters voting in local elections	X	X	X
Percentage reporting trust in city leaders/government	X	X	X
Percentage reporting very good QOL	X		
Percentage perceiving racism as a local problem			X
Racial disparities in juvenile courts	X	X	
Time spent in volunteerism		X	X
Knowing or helping neighbors	X	X	
High school graduation rate	X		X
Percentage of students at or above grade level		X	X
Percentage of licensed child care workers replaced annually		X	
Percent attending artistic or cultural activities	X	X	X
Library circulation	X		X
Average work commute/vehicle miles	X	X	X
Vehicle accidents	X	X	X
Streets with sidewalks	X		
Streets with bike lanes	X		
Direct air flight destinations			X

NOTE: The table shows indicators for which data were available for the three communities shown. Metrics from Austin, Jacksonville, and Seattle were compiled by Greenwood, undated, and verified from their web sites for this report. Because the metrics have changed somewhat since Greenwood's analysis, this table presents summary descriptions of them and does not include all of them, focusing instead on those that are related to QOL as described in this report. It should be noted that other metrics not included could, like many sustainability issues, be related to QOL, such as data about toxic releases and environmental quality.

[a] Percent of households for which average rent would be less than 35 percent of household income.

Bibliography

Andre, P. and D. Bitondo, "Development of a Conceptual and Methodological Framework for the Integrated Assessment of the Impacts of Linear Infrastructure Projects on Quality of Life," Research and Development Monograph Series, 2001. As of September 7, 2007:
http://www.ceaa-acee.gc.ca/015/001/015/title_e.htm

Anniston Army Depot, "Restoration Advisory Board—Fact Sheet," undated. As of December 11, 2006:
http://www.anad.army.mil/restoration/factsheet.shtml

Anniston Army Depot, "Environmental Management System Implementation at Anniston Army Depot," briefing, 8 April 2004. As of September 19, 2007:
www.dtic.mil/ndia/2004enviro/sessions/session13/ems.ppt

Arnold, Matthew B., and Robert M. Day, *The Next Bottom Line: Making Sustainable Development Tangible*, Washington, D.C.: World Resources Institute, 1998.

AtKisson, Alan, *Believing Cassandra: An Optimist Looks at a Pessimist's World*, White River Junction: Chelsea Green, 1999.

Austin, Texas, "Sustainability Indicators Project of Hays, Travis and Williamson Counties," Central Texas Indicators, Austin, Texas, 2000.

Beranek, Dwight A., "Engineer Technical Letter 1110-3-491: Sustainable Design for Military Facilities, appendix B," Washington, D.C., May 1, 2001.

Berkes, F., C. Folke, et al., "Traditional ecological knowledge, biodiversity, resilience and sustainability," in C.A. Perrings, K.G. Maler, C. Folke, C.S. Holling, and B.O. Jansson (eds.), *Biodiversity Conservation*, PO Box 17, 3300 AA Dordrecht, Netherlands, Kluwer Academic Publ. 4, 1995: 281–299.

Besham, Donald L., "Engineering Construction Bulletin: Sustainable Design and Development," unpublished, 2006.

Buntrock, Gerrit, "Food prices: Cheap No More," *The Economist*, December 6, 2007.

Business Roundtable, *Facility Level P2 Benchmarking Study*, Washington, D.C., November 1993.

Business Roundtable, *A Benchmarking Study of Pollution Prevention Planning*, Washington, D.C., August 1998.

Butterfly Conservation Initiative, "BFCI: Taylor's (Whulge) Checkerspot Butterfly," American Zoo and Aquarium Association, 2006. As of September 18, 2007:
http://www.butterflyrecovery.org/species_profiles/taylors_checkerspot/

Camm, Frank, Jeffrey Drezner, Beth E. Lachman, and Susan Resetar, *Implementing Proactive Environmental Management: Lesons Learned from Best Commercial Practice*, Santa Monica, CA:

RAND Corporation, MR-1371-OSD, 2001.
http://www.rand.org/pubs/monograph_reports/MR1371/

Carpenter, B., "Recycling Campaign Highlights Post Efforts for Sustainable Living," *Northwest Guardian*, Fort Lewis, Washington, November 23, 2005.

Carpenter, B., "Fort Lewis to Exceed Army Mandate for Handling Demolition, Construction Debris," *Northwest Guardian*, Fort Lewis, Washington, August 3, 2006.

Cassidy, Robert, "White Paper on Sustainability," *Building Design & Construction*, November 2003, pp. 1–47. As of January 8, 2009:
http://www.bdcnetwork.com/article/CA6326262.html?industryid=42769&q=white+paper+sustainability

Champy, James, *Reengineering Management: The Mandate for New Leadership,* New York: HarperBusiness, 1995.

Costanza, R. and B.C. Patten, "Defining and predicting sustainability," *Ecological Economics* 15(3), 1995: 193–196.

Daily, G.C., and B.H. Walker, "Environmental Sustainability: Seeking the Great Transition." *Nature* 403(6767), 2000: 243.

Estes, R., "Social Development Trends in Europe," *Social Indicators Research*, 42, 1997: 1–19.

Executive Office of the President of the United States, "Executive Order 13123 of June 3, 1999: Greening the Government Through Efficient Energy Management," *Federal Register,* Vol. 64, No. 109, June 8, 1999. As of July 2009:
http://www1.eere.energy.gov/femp/pdfs/eo13123.pdf

Executive Office of the President of the United States, "Strengthening Federal Environmental, Energy, and Transportation Management," Executive Order 13423, *Federal Register*, Vol. 72, No. 17, January 26, 2007. As of July 2009:
http://edocket.access.gpo.gov/2007/pdf/07-374.pdf

Federal Acquisition Regulation (FAR) Subpart 23.4—Use of Recovered Materials and Biobased Products. As of October 2, 2007:
http://acquisition.gov/comp/far/current/html/Subpart%2023_4.html

Federal Acquisition Regulation (FAR) Subpart 23.2—Energy and Water Efficiency and Renewable Energy. As of October 2, 2007:
http://acquisition.gov/comp/far/current/html/Subpart%2023_2.html

Federal Acquisition Regulation (FAR) Subpart 23.7—Contracting for Environmentally Preferable Products and Services. As of October 2, 2007:
http://acquisition.gov/comp/far/current/html/Subpart%2023_7.html

Federal Acquisition Regulation (FAR) Subpart 23.8—Ozone-Depleting Substances. As of October 2, 2007:
http://acquisition.gov/comp/far/current/html/Subpart%2023_8.html

Flynn, P., D. Berry, et al., "Sustainability and Quality of Life Indicators: Toward the Integration of Economic, Social, and Environmental Measures," *Indicators: The Journal of Social Health,* 1(4), 2002.

Fort Benning, "Fort Benning Boot Print: Facts and Figures for the Sustainability Goal-Setting Conference 10–12 May 2005," Fort Benning, Georgia, 2005(a). As of October 20, 2006:
https://www.infantry.army.mil/sustainability/content/9.htm

Fort Benning, "Garrison Commander's Sustainability Message." Fort Benning, Georgia, 2005(b). As of October 20, 2006:
https://www.infantry.army.mil/sustainability/content/1.htm

Fort Benning, "Five Year Strategic Plan (Draft)," Fort Benning, Georgia, 15 February 2007.

Fort Bliss, "Mission and Master Plan: Programmatic Environmental Impact Statement." Fort Bliss, Texas, December 2000. As of October 27, 2006:
https://www.bliss.army.mil/About%20Ft%20Bliss/NEW-EIS/index_EIS.htm

Fort Bliss, "Fort Bliss Integrated Natural Resources Management Plan." Fort Bliss, Texas, November 2001. As of October 27, 2006:
https://www.bliss.army.mil/About%20Ft%20Bliss/NEW-EIS/index_EIS.htm

Fort Bragg, "Installation Design Guide for a Sustainable Fort Bragg," Fort Bragg, North Carolina, undated. As of September 18, 2007:
http://www.bragg.army.mil/dpw/idg

Fort Bragg, "Fort Bragg/Pope AFB Regional Land Use Advisory Commission, "Living in Harmony with Fort Bragg and Pope AFB." Fort Bragg/Pope Air Force Base Joint Land Use Study, May 2003(a).

Fort Bragg, "Sustainable Fort Bragg," newsletter, Fort Bragg, NC, Volume 3, December 2003(b).

Fort Bragg, "Sustainable Fort Bragg: Annual Report." Fort Bragg, NC, May 2003(c).

Fort Bragg, "Sustainable Fort Bragg: Success Stories." Fort Bragg, NC, 2003(d). As of December 18, 2007:
http://www.bragg.army.mil/Sustainability/successStories.htm

Fort Bragg, "Master Plan: Long Range Component Fort Bragg, North Carolina," Fort Bragg, NC, Public Works Business Center, February 2004(a).

Fort Bragg, "Sustainable Fort Bragg," newsletter, Fort Bragg, NC, Volume 4, December 2004(b).

Fort Bragg, "Sustainable Fort Bragg," newsletter, Fort Bragg, NC, Volume 5, June 2005.

Fort Bragg, "Fort Bragg Strategic Campaign Plan FY07-FY11," Fort Bragg, NC, June 2006.

Fort Bragg, "Sustainable Fort Bragg: Goal 5: Compatible Land Use Objectives," May 8, 2006. As of January 8, 2009:
http://www.bragg.army.mil/Sustainability/goal5-objs.htm

Fort Bragg, "Strategic Garrison Goal 1 (briefing)." Fort Bragg, NC, Department of Public Works, April 12, 2007(a).

Fort Bragg, "Sustainable Fort Bragg (newsletter) Volume 10." Fort Bragg, NC, December 2007(b).

Fort Campbell, "Army Installation Success Stories: Demolition Concrete Waste Diversion," fact sheet, Fort Campbell, KY, undated (a).

Fort Campbell, "Installation Strategic Plan: Sustainability Program Annex, Implementation Plan 2005–2012." Fort Campbell, KY, undated (b). Received from Fort Campbell August 2007.

Fort Campbell, "Strategic Plan 2007–2012." Fort Campbell, KY, undated (c). Received from Fort Campbell August 2007.

Fort Campbell, "Installation Sustainability Baseline," Fort Campbell, KY, September 2003. Received from Fort Campbell August 2007.

Fort Campbell, "Sustainable Installation Management System ISO 14001." Fort Campbell, KY, 15 January 2006. As of October 19, 2006:
www.campbell.army.mil/envdiv/SIMS250_MONITORING_AND_MEASUREMENT.pdf

Fort Carson, "Secretary of the Army 2007 Sustainability Award: Installation Nomination: Fort Carson, Colo.," memo, undated.

Fort Carson, "Fort Carson Sustainability Baseline," Fort Carson, CO, August 2002.

Fort Carson, "Fort Carson 2005 Sustainability Progress Report." Fort Carson, CO, 2005.

Fort Carson, "Fort Carson 2006 Sustainability Report," Fort Carson, CO, 2006.

Fort Carson, "Fort Carson 2007 Sustainability Report," Fort Carson, CO, 2007.

Fort Eustis, "Fort Eustis and Fort Story Environmental Management Handbook," Fort Eustis, VA, August 2003. As of October 26, 2006:
http://dpw-web.eustis.army.mil/enrd/environmental_documents.htm

Fort Hood, "Fort Hood Sustainability Progress Report Fall 2003." Fort Hood, TX, 2003. As of October 20, 2006:
www.sustainability.army.mil/resources/libdocs_activity/fort_hood/Hood%20Progress%20 Report%20Fall%202003.pdf

Fort Hood, "Fort Hood 2005 Sustainability Annual Progress Report." Fort Hood, TX, 2005(a). As of October 26, 2006:
http://www.sustainability.army.mil/resources/library_activity.cfm

Fort Hood, "Sustainability at Fort Hood." Fort Hood, TX, 2005(b).

Fort Hood, "Solid Waste Annual Report for FY06," Fort Hood, TX, 2006.

Fort Lewis, "Fort Lewis Executive Summary, Five Year Plan: Sustainability Implementation Plan," Fort Lewis, WA, 2003(a).

Fort Lewis, "Fort Lewis Sustainability Annual Report April 2003," Fort Lewis, WA, 2003(b).

Fort Lewis, "Memorandum: Sustainable Facility, Construction, and Operations Policy," Headquarters, I Corps and Fort Lewis, WA, 2003(c). As of October 18, 2006:
http://www.lewis.army.mil/publicworks/

Fort Lewis, "Sustainability Annual Report April 2003," Fort Lewis, WA, 2003(d). As of October 18, 2006:
http://www.lewis.army.mil/publicworks/

Fort Lewis, "Sustainable Fort Lewis: Progress Report for April 2003–April 2004," Fort Lewis, WA, 2004.

Fort Lewis, "Sustainable Fort Lewis: April 2004–April 2005 Progress Report," Fort Lewis, WA, 2005. As of October 18, 2006:
http://www.lewis.army.mil/publicworks/

Fort Lewis, "Sustainable Fort Lewis: 2005–2006 Annual Report," Fort Lewis, WA, 2006.

Fort Lewis, "Fort Lewis Executive Summary Five Year Plan: Sustainability Implementation Plan," Fort Lewis, WA, undated (a). As of October 18, 2006:
http://www.lewis.army.mil/publicworks/

Fort Lewis, "Sustainability Homepage FLW Installation Sustainability Program (ISP)," Fort Lewis, WA, undated (b). As of September 18, 2007:
http://www.lewis.army.mil/publicworks/

Fort Lewis, "The Sustainable Interiors Showroom (factsheet)," Fort Lewis, WA, undated (c).

Fort Monroe, "Garrison Strategic Plan 2006–2011," Fort Monroe, VA, undated. As of October 26, 2006: 147.248.251.93/monroe/sites/local/StrategicPlan2006thru2011.pdf

Fort Monroe, "Fort Monroe EMS Implementation Plan." Fort Monroe, VA, 27 September 2004. As of October 26, 2006:
www.tradoc.army.mil/dpw/Fort%20Monroe%20EMS%20Implementation%20Plan.doc

Fort Monroe, "FY 2006 Fort Monroe Installation Action Plan," Fort Monroe, VA, July 2005. As of October 26, 2006:
http://monroeapps.monroe.army.mil/Fort_Monroe/sites/directorates/dpw/dpw.aspx

Fort Polk, "Fort Polk Environmental Management System," Fort Polk, LA, 8 February 2006. As of October 26, 2006:
www.sustainability.army.mil/tools/docs_ems/EMS%20Web%20Page%20Items/Fort_Polk_EMS_Brochure.pdf

Fort Rucker, "Fort Rucker's Environmental Management System: Understanding Your Role," U.S. Army Aviation Center and Fort Rucker, AL, undated. As of October 27, 2006:
http://www.fortrucker-env.com/documents.aspx?docID=36&cur=25

Fort Rucker, "Fort Rucker Environmental Management System," U.S. Army Aviation Center and Fort Rucker, AL, 2005. As of October 26, 2006:
http://www.fortrucker-env.com/documents.aspx?docID=36&cur=25

Fort Stewart, "Fort Stewart/Hunter Army Airfield Sustainability Management System Manual," Fort Stewart, GA, January 2006.

Galentine, S., "Fort Carson Forges Strong Regional Partnerships," undated (a). As of September 25, 2007:
http://www.susainability.army.mil/news/success.cfm

Galentine, S., "Fort Carson Invests In Renewable Energy Production," undated (b). As of September 25, 2007:
http://www.sustainability.army.mil/news/success.cfm

Galentine, S., "MEDDAC Soldier's New Barracks Provide Creature Comfort in an Eco-Friendly Way," U.S. Army, undated (c). As of September 25, 2007:
http://www.sustainability.army.mil/news/success.cfm

Galentine, S., "Pollution Prevention Projects Help Increase Motorpool Safety," undated (d). As of September 25, 2007:
http://www.sustainability.army.mil/news/success.cfm

Galentine, S., "Sustainability Initiatives Reduce Hazardous Air Pollutants," undated (e). As of September 25, 2007:
http://www.sustainability.army.mil/news/success.cfm

Glaser, C. and A. Cohen, "Fort Hood Army Base Environmentally Friendly," Broadcast: National Public Radio, Marketplace Program, February 8, 2006.

Global Environmental Management Initiative (GEMI), home page. As of January 8, 2009:
http://www.gemi.org/gemihome.aspx

Global Environmental Management Initiative (GEMI), *Benchmarking, the Primer: Benchmarking for Continuous Environmental Improvement*, Washington, D.C., 1994.

Global Environmental Management Initiative (GEMI), "GEMI SD Planner & SD Gateway Use Manual," V 3.1, October 25, 2006.

Global Environmental Management Initiative (GEMI), "The Metrics Navigator," March 2007. As of January 8, 2009:
http://www.gemi.org/metricsnavigator/

Goodland, R., "The Concept of Environmental Sustainability," *Annual Review of Ecology and Systematics* 26, 1995: 1–24.

Greenwood, D., "Local Indicators of Quality of Life: A Preliminary Look at the Pikes Peak Region Introduction." Colorado: Center for Colorado Policy Studies, undated.

Greenwood, D., "Measuring Quality of Life with Local Indicators," Presentation: Jerome Levy Institute Conference on Quality of Life in Industrialized Countries, 2001. As of September 10, 2007:
http://web.uccs.edu/ccps

Grumbine, R.E., "What Is Ecosystem Management?" *Conservation Biology*, 8(1), 1994.

Hadekel, Peter, "Green Building Movement Is Growing Some Legs," *The Gazette*, Montreal, April 6, 2007. As of December 18, 2007:
http://www.iliv.com/company/news/hadekel.cfm

Harris, R., "LEED Takes Hold at Fort Bragg," *Public Works Digest* May/June 2007.

Hawken, Paul, Amory Lovins, et al., *Natural Capitalism: Creating the Next Industrial Revolution,* Boston, MA: Little, Brown and Company, 1999.

International City/County Management Association (ICMA), *Getting to Smart Growth: 100 Policies for Implementation*, 2002.

Jackson, Margaret, "Beyond the PX," *The Denver Post*, May 31, 2007. As of December 18, 2007:
http://www.denverpost.com/business/ci_6023851

Jacksonville Council, "Quality of Life in Jacksonville: Indicators for Progress." Jacksonville, Florida Community Council, 2001. As of September 8, 2007:
http://web.jcci.org/quality_of_life/quality_of_life.pdf

Jenicek, Elisabeth M., Natalie R.D. Myers, Brad Boesdorfer, and Donald F. Fournier, "Strategic Sustainability Assessment Pilot Study: Fall Line Region of the Southeast," USACE, ERDC/CERL TR-06-32, November 2006.

Kotter, John P., *Leading Change*, Boston: Harvard Business School Press, 1996.

Kotter, John P., et al., *Our Iceberg Is Melting: Changing and Succeeding Under Any Conditions,* New York: St. Martin's Press, 2006.

Kramer, D., "Solar Wall Conserves Energy," *Northwest Guardian*, Fort Lewis, WA, May 18, 2007.

Lachman, Beth E., *Linking Sustainable Community Activities to Pollution Prevention: A Sourcebook,* Santa Monica, CA: RAND Corporation, MR-855-OSTP, 1997.
http://www.rand.org/pubs/monograph_reports/MR855/

Lachman, Beth E., et al., *Integrated Facility Environmental Management Approaches: Lessons from Industry for Department of Defense Facilities*, Santa Monica, CA: RAND Corporation, MR-1343-OSD, 2001.
http://www.rand.org/pubs/monograph_reports/MR1343/

Lachman, Beth E., et al., *The Thin Green Line: An Assessment of DoD's Readiness and Environmental Protection Initiative to Buffer Installation Encroachment*, Santa Monica, CA: RAND Corporation, MG-612-OSD, 2007.
http://www.rand.org/pubs/monographs/MG612/

Luciano, Christine, "Going Lean, Clean, and (Army) Green: DPW Tackles Environmental Issues," Fort Hood Department of Public Works Environmental News Release, Fort Hood, TX, July 2007. As of September 19, 2007:
http://www.dpw.hood.army.mil/Files/Engineers%20go%20green.pdf

Miller, K., "Army Environmental Programs Awarded for Making a Difference," *U.S. Fed News*, June 13, 2007.

Mueller, M., "Community Indicators for Denver." Denver: Center for Colorado Policy Studies (Contributed Paper #1), 1999. As of September 9, 2007:
http://web.uccs.edu/ccps

Nattrass, Brian, and Mary Altomare, *Dancing with the Tiger: Learning Sustainability Step by Natural Step*, Gabriola Island, BC, Canada: New Society Publishers, June 2002.

Nature Conservancy, "Rare Butterfly Emerges on the Puget Sound Prairie," 2007. As of September 18, 2007:
http://www.nature.org/wherewework/northamerica/states/washington/preserves/art21325.html

Neely, B., S. Kettler, J. Horsman, C. Pague, R. Rondeau, R. Smith, L. Grunau, P. Comer, G. Belew, F. Pusateri, B. Rosenlund, D. Runner, K. Sochi, J. Sovell, D. Anderson, T. Jackson, and M. Klavetter, *Central Shortgrass Prairie Ecoregional Assessment and Partnership Initiative,* The Nature Conservancy of Colorado and the Shortgrass Prairie Partnership, Final Report, November 2006. As of July 2009:
http://www.cnhp.colostate.edu/documents/2006/CSP%20Final%20Report%202006.pdf

New Zealand, "Quality of Life in New Zealand's Eight Largest Cities," 2003. As of September 19, 2007:
http://www.bigcities.govt.nz/pdf/Quality_of_Life_03.pdf

Norton, Bryon G., *Toward Unity Among Environmentalists*, New York: Oxford University Press, 1991.

Norton, Bryon G., *Sustainability: a Philosophy of Adaptive Ecosystem Management*, Chicago: University of Chicago Press, 2005.

Phillips, D., "Quality of Life and Sustainability," *The International Journal of Environmental, Cultural, Economic and Social Sustainability*, 2(2), 2006: 103–112.

Pikes Peak Sustainability Indicator Project (PPSIP), "Project Summary Report (Final Draft)," Fort Carson, CO, 2005.

Porter, Douglas R., *Managing Growth in America's Communities*, Washington, D.C., Island Press, 1997.

PR Newswire, "Fort Lewis on the Way to Zero Net Waste Generation," *U.S. Newswire*, June 3, 2007.

Putnam, R., "The Prosperous Community: Social Capital and Public Life." *The American Prospect*, Spring, 1993: 35–42.

Rawlins, Wade, "Engineer's Work Grows Greener Army: Fort Bragg Leader Pushes for Earth-Friendly Post," *The News & Observer*, Fayetteville, NC, November 23, 2007. As of December 18, 2007:
http://www.newsobserver.com/news/story/786616.html

Sandhills Association, "Joint Land Use Study (JLUS)," Sandhills GIS Association, undated. As of September 18, 2007:
http://sandhillsgis.jlus.htm

Schafer, M., B. Nolting, et al., "Bringing Together the Concepts of Quality of Life and Sustainability," in W. Glatzer, S. von Below, and M. Stoffregen (eds.), *Challenges for Quality of Life in the Contemporary World,* Social Indicator Research Series Vol. 24, Dordbrecht/Boston/London: Kluwer, 2004, pp. 33–43.

Schulz, William, "The Cost of Biofuels," *Chemical & Engineering News*, December 17, 2007.

Shackman, G., Y.-L. Liu, et al., "Measuring Quality of Life Using Free or Public Domain Data," Social Research Update, 2005. As of September 11, 2007:
http://www.soc.surrey.ac.uk/sru/

Snyder, N., "Fort Lewis Takes Creative Roads to Sustainability," U.S. Army Environmental Center, 2005.

Snyder, N., "Building an Iraqi Village; Focus on the Environment," *Soldiers Magazine* 61(4), 2006: 12–13.

Soldiers Magazine, "Smarter Building Recycling," 62(4), 2007: 34–38.

Sustainable Sandhills, "Our Mission and Our Vision," web site. As of January 7, 2009:
http://www.sustainablesandhills.org/

Sustainable Seattle, "Indicators of Sustainable Community," 2004, reprinted from 1998. No new updates were printed since the original. As of September 9, 2007:
http://www.scn.org

Unisys Corporation, "Base Operations Assessment," Washington, D.C., Prepared for the Office of the Deputy Under Secretary of Defense (Installations and Environment), Installation Requirements and Management, April 28, 2005. As of October 3, 2007:
http://www.acq.osd.mil/ie/irm/irm_library/BOA%20Final%20w%20Apdx.pdf.

United Nations, "Human Development Report," United Nations Development Programme, 2004. As of September 13, 2007:
http://hdr.undp.org/reports/global/2004/

U.S. Army, "Alternative Fueled Vehicles Part of Fort Carson Fleet," undated (a). As of September 24, 2007:
http://www.sustainability.army.mil/news/successStory06-01.cfm

U.S. Army, "Deconstruction Puts New Twist on Removing Old Buildings," undated (b). As of September 25, 2007:
http://www.sustainability.army.mil/news/success.cfm

U.S. Army, "Journey to Sustainability: U.S. Army Garrison Hawaii," undated (c). As of September 24, 2007:
http://www.sustainability.army.mil/news/successStory06-06.cfm

U.S. Army, "Purchase of Alternatively Fueled Vehicles and E85 Fueling Points," undated (d). As of September 25, 2007:
http://www.sustainability.army.mil/news/successStory06-02.cfm

U.S. Army, "Army Energy Program," undated (e). As of January 8, 2009:
http://army-energy.hqda.pentagon.mil/policies/epa.asp

U.S. Army, Field Manual 100-22, *Installation Management,* Washington, D.C., Headquarters, Department of the Army, 1994

U.S. Army, "The Army Strategy for the Environment: Sustain the Mission—Secure the Future." Washington, D.C., U.S. Army Environmental Policy Institute, 2004. As of October 2, 2007:
https://www.asaie.army.mil/Public/ESOH/doc/ArmyEnvStrategy.pdf

U.S. Army, "Army Regulation 350-19: The Army Sustainable Range Program," Washington, D.C., Headquarters, Department of the Army, 30 August 2005.

U.S. Army, "Command and Control Facilities: Division Through Army Command," Washington, D.C., U.S. Army Corps of Engineers Savannah District, 2007(a).

U.S. Army, "Directorate of Military Programs Weekly Update: Sustainable Design and Development," Washington, D.C., U.S. Army Corps of Engineers, October 12, 2007(b).

U.S. Army, "The Strategic Plan for Army Sustainability," unpublished draft, Headquarters, Department of the Army, Washington, D.C., July 25, 2007(c).

U.S. Army, "Army Environmental Programs Awarded for Making a Difference," *States News Service*, Washington, D.C., June 13, 2007(d).

U.S. Army, Army Regulation 200-1, "Environmental Protection and Enhancement," Headquarters, Department of the Army, December 13, 2007(e). As of June 2009:
http://www.army.mil/usapa/epubs/pdf/r200_1.pdf

U.S. Army Corps of Engineers (USACE) Engineering Construction Bulletin (ECB 2003-20 SPiRiT), November 7, 2003 (revised 10 May 2006).

U.S. Army Corps of Engineers (USACE), "Directorate of Military Programs Weekly Update: Sustainable Design and Development (SDD)," October 12, 2007.

U.S. Army Installation Management Command Southeast Region, "The Guide to Creating a Sustainable Installation," 2006.

U.S.–Asia Environmental Partnership, *Global Environmental Management: Candid Views of Fortune 500 Companies*, Washington, D.C. October 1997.

U.S. Fed News, "Sustainability Summit: Securing Fort Jackson's Future," *U.S. Fed News,* March 10, 2006.

U.S. Green Building Council, "Leadership in Energy and Environmental Design," undated. As of January 8, 2009:
http://www.usgbc.org/DisplayPage.aspx?CategoryID=19

U.S. Green Building Council, "Green Building Rating System for New Construction & Major Renovations, Version 2.2," October 2005.

U.S. Green Building Council, "Green Building, USGBC and LEED," 2007(a). As of November 28, 2007:
http://www.usgbc.org

U.S. Green Building Council, "LEED for Existing Buildings," 2007(b). As of November 28, 2007:
http://www.usgbc.org/DisplayPage.aspx?CMSPageID=221

U.S. Green Building Council, "LEED for New Construction," 2007(c). As of November 28, 2007:
http://www.usgbc.org/DisplayPage.aspx?CMSPageID=220#v2.2

U.S. Green Building Council, "Registration," 2007(d). As of November 28, 2007:
http://www.usgbc.org/DisplayPage.aspx?CMSPageID=65

U.S. Green Building Council, "US Green Building Council Home Page," 2007(e). As of November 28, 2007:
http://www.usgbc.org/DisplayPage.aspx?CMSPageID=124

U.S. Senate, "Green Buildings: Statement of Peter Templeton, Vice President, Education and Research." Washington, D.C., United States Senate, USGBC hearing before the Committee on Senate Environment and Public Works, May 15, 2007.

World Business Council for Sustainable Development, web site. As of January 8, 2009:
http://www.wbcsd.org/

World Business Council for Sustainable Development, "Information and Publications," brochure, Geneva, Switzerland, undated. As of September 29, 2008:
http://www.wbcsd.ch/

World Commission on Environment and Development, *Our Common Future,* Oxford: Oxford University Press, 1987.